Basic Molecular Protocols in Neuroscience: Tips, Tricks, and Pitfalls

Basic Molecular Protocols in Neuroscience: Tips, Tricks, and Pitfalls

John T. Corthell, Ph.D.

AMSTERDAM • BOSTON • HEIDELBERG • LONDON
NEW YORK • OXFORD • PARIS • SAN DIEGO
SAN FRANCISCO • SINGAPORE • SYDNEY • TOKYO

Academic Press is an imprint of Elsevier

Academic Press is an imprint of Elsevier
32 Jamestown Road, London NW1 7BY, UK
The Boulevard, Langford Lane, Kidlington, Oxford OX5 1GB, UK
Radarweg 29, PO Box 211, 1000 AE Amsterdam, The Netherlands
225 Wyman Street, Waltham, MA 02451, USA
525 B Street, Suite 1800, San Diego, CA 92101-4495, USA

First published 2014

Library of Congress Cataloging-in-Publication Data
A catalog record for this book is available from the Library of Congress

British Library Cataloguing-in-Publication Data
A catalogue record for this book is available from the British Library

ISBN: 978-0-12-801461-5

For information on all Academic press publications
visit our website at store.elsevier.com

This book has been manufactured using Print On Demand technology. Each copy is produced
to order and is limited to black ink. The online version of this book will show color figures
where appropriate.

Academic Press is an imprint of Elsevier
32 Jamestown Road, London NW1 7BY, UK
The Boulevard, Langford Lane, Kidlington, Oxford OX5 1GB, UK
Radarweg 29, PO Box 211, 1000 AE Amsterdam, The Netherlands
225 Wyman Street, Waltham, MA 02451, USA
525 B Street, Suite 1800, San Diego, CA 92101-4495, USA

First published 2014

Library of Congress Cataloging-in-Publication Data
A catalog record for this book is available from the Library of Congress

British Library Cataloguing-in-Publication Data
A catalogue record for this book is available from the British Library

ISBN: 978-0-12-801461-5

For information on all Academic press publications
visit our website at store.elsevier.com

This book has been manufactured using Print On Demand technology. Each copy is produced
to order and is limited to black ink. The online version of this book will show color figures
where appropriate.

Working together
to grow libraries in
developing countries

www.elsevier.com • www.bookaid.org

CONTENTS

INTRODUCTION

My goal for this book is to share my practical, technical knowledge with you to save you years of hardship with experiments that refuse to work because you haven't fine-tuned the techniques. This book contains protocols, solution recipes, ideas for experimental controls and trouble-shooting, and descriptions of what each technique really is—what you are doing at each step and why you are doing it—for basic neurobiology protocols. I did my best to include tips to increase the efficiency of these protocols and techniques, as well as discussing different issues that could compromise the experiment (no one thinks about hydrogen peroxide degradation until nothing works). These techniques may also be called "molecular neurobiology," which is simply "molecular biology" in a neuroscience context, in case you were wondering what you were really getting here.

Please note that these protocols are not attributed to any one specific person. It doesn't mean that I came up with the techniques myself; it means that these techniques are very common and you can find hundreds of iterations via the Internet or the laboratories you interact with, so these are Franken-protocols (bolts in the neck and all) that I cobbled together from what I considered to be the best practices and protocols that I have seen. I take credit only for my own experiences and tips that I give you here.

I have had many teachers, whether professors, graduate students, technicians, undergraduates, company employees, or my own failures. These teachers have been invaluable to garnering the experience that I write about in this book, whether we had a formal meeting, I took a course, or just had conversations in the hallways (the majority of my education has been in hallways and in failures).

Many thanks to the following:

P. Trombley, J. Olcese, C. Levenson, G. O'Neal-Moffitt, A. Irsigler, S. Miller, B. Washburn, C. Fitch-Pye, C. Helena, B. Kwon, T. Houpt, C. Ouimet, and B.D. Lynn. Thanks to P. Trombley and L. Biggs for reviewing the manuscript. To any that I have forgotten, I extend my apologies to them for a faulty memory.

> *The working definition of an expert is someone who has failed enough times to tell you what not to do.*
>
> **Unknown**

> *Do not be discouraged by failure, for every wrong attempt discarded is another step forward.*
>
> **Unknown**

CHAPTER 1

General Notes

Basic molecular protocols require a basic understanding of solutions chemistry (i.e., the concepts of molarity, molality, pH, and stoichiometry). If you do not know how to calculate molarity from weight (in grams) and volume, or how to calculate the grams you need for your solution from a molarity value, then please review a basic chemistry text. Otherwise, some of these instructions will be incomprehensible.

Typically, solutions are stored and labeled at some molarity concentration (M, mM, μM, etc.), but many protocols and recipes refer to a multiplier value, such as $10\times$, $5\times$, or $1\times$. The "$\times$" is a multiplier and tells you how much you'll need to dilute that stock solution. Unless the protocol says otherwise, you will generally use solutions at a $1\times$ concentration. For example, if I have a $10\times$ stock solution and I want 1 liter of $1\times$ working solution, I will dilute 100 milliliters (ml) of $10\times$ stock with 900 ml of water to make my $1\times$ working solution (i.e., divide a liter into 10 parts to find the amount of stock solution to use).

Remember your metric prefixes. Using molarity (M) as an example: millimolar is 10^{-3} molar (mM), micromolar is 10^{-6} molar (μM), nanomolar is 10^{-9} molar (nM), and picomolar is 10^{-12} molar (pM).

"Multiplexing" means that you run multiple experiments in a single tube or tissue sample (or similar reaction site). Multiplexing a quantitative polymerase chain reaction (PCR) (see Chapter 4) means that, using unique probes for each target, you amplify multiple targets within a single tube or well. In immunoblotting and immunohistochemistry, multiplexing would mean that you use two or more antibodies on a single blot/tissue slice at the same time and visualize them at the same time (possible via fluorescence and some colorimetric reactions). In *in situ* hybridization, multiplexing means using two or more probes in the tissue at the same time and visualize them at the same time (again, via fluorescence and some colorimetric reactions). Multiplex reactions are great if you can get them, but be aware that the optimal conditions for one reaction may be the worst conditions for the other reaction. Additionally, if you have a limited amount of

Basic Molecular Protocols in Neuroscience: Tips, Tricks, and Pitfalls. DOI: http://dx.doi.org/10.1016/B978-0-12-801461-5.00001-0

one crucial reagent, such as deoxynucleotide triphosphates (dNTPs), that limiting reagent will be used for all of your reactions simultaneously and each reaction will, therefore, affect the others.

"Vortexing" means that you use a device called a vortexer (we scientists are a creative lot) until you see the "cyclone" in the center of the tube. This can rapidly mix solutions, but it is inappropriate if your solution components are sensitive to mechanical forces. For example, DNA is sensitive to mechanical force and solutions containing DNA should not be mixed via vortexer.

"Pipetting" means that you are using pipets to measure some volume of whatever solutions you are using in your experiments. Pipetting is a broad term, encompassing the use of rubber balls on the ends of labeled pipets, micropipettors and disposable tips, disposable bulb pipets, and labeled pipets with hand-held electrical pumps (called pipette guns). When in doubt, pipette solutions in and out slowly and make sure the liquid is all out or nearly so. When changing the volume measurement on the micropipettor, perform it exactly how the manufacturer stipulates—if they have a wheel in the middle, then using the top will eventually take the top off! If they don't have a wheel, then turn the top of the micropipettor.

Develop your pipetting technique. Many of these experiments depend on your ability to accurately pipette the correct amount of fluid. You should pipette solutions slowly and evenly for best results. After much practice, you will be able to quickly pipette accurate amounts of solution. Additionally, it is common and necessary to mix solutions by pipetting the fluid in and out of the tip repeatedly. Nucleotides, among other things, like to stick to plastic, and pipetting back and forth brings them back into solution. If you did not mix thoroughly, the amount of liquid you pipette might be correct, but the things inside the liquid that you want may not be at the correct concentration. I prefer to count my passes (one pass = into the pipette tip and back into the container), and consider 30 passes to be mixed sufficiently. Finally, plan your pipetting to a minimum number of steps. Every time you pipet, you increase the risk of experimenter-based errors and contamination. For example, if I plan to pipette small volumes three times, I have fewer chances to contaminate my solutions than if I pipette smaller volumes six times. Even better, use a different micropipettor and pipette once.

When adjusting the pH of Tris-based solutions, use a Tris electrode-based pH meter. When adjusting the pH of paraformaldehyde (PFA) solutions, use disposable pH strips (most pH meter electrodes aren't tested against PFA solutions, so they may damage the electrode). For other solutions, a standard pH meter is fine. Always continuously stir solutions (either by hand or by stir bar) while adjusting the pH, or else you may measure the pH of your acid or base instead of the solution as a whole. Follow manufacturer instructions for proper storage of your pH electrode. If you do not know how to use pH solutions, the manufacturers of pH meters have instructions included with their products or on their web sites. Additionally, you should be able to ask someone in the lab if those instructions don't work.

A freeze-thaw cycle is, any time where a frozen solution is thawed, used, then frozen again. Just about anything that you want to freeze is sensitive to the number of times it is thawed. This is the big reason that you should use aliquot solutions (see Aliquoting section) and avoid frost-free freezers. Frost-free freezers avoid forming frost by essentially thawing at some critical point (a specific temperature, time, or other factor), which means that your samples in a frost-free freezer will go through many freeze-thaw cycles without you knowing it. Repeated freeze-thaw cycles will render your antibodies useless, degrade guanosine-5′-triphosphate (GTP), and degrade nucleotides.

When troubleshooting, a "known good" has immense value. A "known good" is something that you know works: a wire that you've tested and conducts electricity as it should, an antibody that gives great results every time, a PCR primer set that is specific, or a selection of tissue that has generated great results in the past. Being able to test a known good against your new, unknown results can eliminate a number of variables from your troubleshooting (if your known good antibody doesn't generate staining, for example, then the problem is not necessarily the antibody but could be the signal-generating components or your sample preparation).

"Common use equipment" is typically a euphemism for "no one takes care of this machine." Even if someone does take care of the equipment (you should be grateful if this is the case), it will still benefit you greatly to read the user's manual and, if necessary, perform routine maintenance days or hours before you use the machine (this is a must if no one takes care of the machine). By performing maintenance before your experiment, you lower the risk of having the machine break down at a crucial point and ruining your experiments. Finally, clean up after yourself when using a common-use machine. No one needs to examine a spill in a common centrifuge and wonder if that spill is safe or contains a deadly virus.

Some abbreviations are used in multiple places and mean different things. You may find a lab notebook that refers to RT as reverse transcription, room temperature, or real time. While I will avoid that in this book, always look for context clues as to which abbreviation means what.

If you are borrowing equipment or space from anyone else, remember: leave a place better than you found it. If you happen upon a lab with impeccably clean and organized shelves and perfect conditions, then do two things. First, take a picture for posterity, and second, leave the lab in the same impeccable condition that you found it.

ASEPTIC TECHNIQUE

Before starting any of these molecular biology experiments, it is important to understand and be willing to refine your aseptic technique. As I understand it, "aseptic technique" is a phrase from microbiology but is perfectly applicable to molecular biology. Proper aseptic technique avoids introducing any unwanted materials (bacteria, molds, or unwanted solutions) into your reactions.

A very important factor in aseptic technique is sterilizing solutions. There are many ways to sterilize solutions, but only a few are important in practice: autoclaving, filter sterilization, and bleach. Bleach is used when you have a dangerous material to dispose off and you need to be absolutely certain that material is absolutely dead before it goes into the trash (where your unsuspecting janitorial staff could be hurt if you are negligent). Autoclaving is using a combination of heat and pressure to kill bacteria and destroy bacterial endospores. Autoclaving takes place in a specially designed machine called (ta-da!) an autoclave. Autoclaves have multiple settings and can vary by manufacturer. One uses autoclave tape to seal or mark items for autoclaving; autoclave tape is designed to have specific patterns after autoclaving to signal that your item is sterilized. In general, 5−20 min of autoclaving is enough for plastic tips and tubes (check the melting temperature by manufacturer to make sure your plastics won't melt in the autoclave). Autoclave-safe bottles need to have lids on loosely. Try not to mix biohazardous waste with nonbiohazardous materials in the autoclave, as anything dangerous needs more time in the autoclave for people to be absolutely certain that it is no longer a threat. Filter sterilization is for solutions that contain vitamins or components that break down due to the heat and pressure of autoclaving. Check every component of your solutions to make sure that you can autoclave them and still have the intended solution.

Other than starting with clean solutions, using a great deal of proper aseptic technique is common sense. If your pipette tip just fell onto the lab bench, you do not use that tip for your experiment; you throw it away immediately. If you do use that tip, you won't know if the experiment failed because that's the true result, or if it was because you introduced bacteria that treated your samples as food. If the experiment works, was it because your experiment was a success or did you get a false positive from that mold you inadvertently introduced? I knew someone who thought for 6 months that they had discovered a completely new protein in their samples ("I'll be famous!"), only to discover that they contaminated their samples with their own skin cells for 6 months. Developing your aseptic technique is very important to avoid wasting your time and funds. Here are some ideas to get you started:

1. *Wear gloves. Always wear disposable gloves and relevant personal protective equipment.* You wouldn't believe how many people don't use gloves and wonder why their samples look funky (it's a technical term). Remember that "molecular biology" means "wear gloves." You wear gloves in some cases to protect yourself, but here you wear gloves to protect your samples from you and your bacteria. If you are working with pathogenic bacteria or viruses, then you are protecting yourself and protecting your samples. Remember that working without gloves while experimenting on dangerous materials (Ebola virus, HIV) will only hurt you. If necessary, gloves can be autoclaved in commercially available autoclave bags or wrapped in paper towels then aluminum foil and sealed with autoclave tape.

2. *Only use autoclaved pipette tips.* Some manufacturers sell RNase-, DNase-, and pyrogenase-free tips. These tips are fine to use; you just want to make sure that your tips have been sterilized or cleaned up in some way (aerosol-free tips help with pipetting but don't

affect aseptic properties). Autoclaving doesn't necessarily get rid of RNases (nucleases that destroy RNA and may therefore destroy your samples), but the plastics can't handle the multiple hours at 200°C that are guaranteed to destroy RNases. In practice, autoclaving works just fine for tips and plastic sample tubes. Tubes and tips that are in large bags are exposed to bacteria and other things as soon as you open the bag, so if you purchase those items in bulk bags, autoclave them in separate containers to ensure their cleanliness. If you drop your tips on the floor or a lab bench, don't use them. If you are really cheap, they can be washed and reautoclaved (people do this), but don't ruin your experiment by using dirty tips. A tip costs less than 10 cents (USD); what is that compared to the cost of the rest of your experiment?

3. *Autoclave or otherwise clean your tools as necessary.* It should go without saying that any tools used for surgery or dissection need to be washed after use. Tools can be autoclaved in the same manner as gloves, though scissors and other sharp implements may require resharpening after autoclaving. Tools can also be washed with 70% ethanol (this percentage works best for killing bacteria, compared to 100% ethanol or 50% ethanol solutions) and/or commercially available RNase-destroying solutions. Some people dip their tools directly into RNase inhibitors between samples.

4. *Autoclave or otherwise sterilize all solutions that you will use for PCR.* Most solutions are sold as sterile or cleaned solutions; these are fine unless you have some reason not to trust them. However, if you do not trust the solution or you made it from scratch and used nonsterile components, sterilize the solutions. Sterile solutions are a necessity for PCR, electrophysiology, immunoblotting, immunohistochemistry, and for long-term sample storage (check Solutions in those sections to see about which solutions are sterilized and which are not).

5. *Clean up the working area before each use.* If you can, designate a single area for each type of experiment (a PCR setup area, separate from where your reaction occurs). If you don't have a designated area for performing these experiments, clean up the bench with 70% ethanol. If you are working with RNA, you can also use RNase-destroying solutions to clean your workspace.

6. *Perform work in a "known good" clean area.* This relates to the previous tip. If you can, having a designated space that is kept clean (a dedicated sterile hood, for example) can be quite beneficial.

The sterile hood keeps particles out via HEPA filters, while you keep the bottom clean by washing with 70% ethanol before and after each use. I keep sterilized pipettors, pipette tips, and tubes in the dedicated hood to avoid introducing new contaminants on a regular basis. Common-use tips, tubes, and pipettors are a small price to pay for people not introducing new molds, bacteria, and junk into your sterile hood by bringing their equipment with them every time they want to run an experiment.

7. *If your tip touched something that is not sterile, throw it out and don't introduce it to your experimental solution.* Let's say you grabbed a capped tube with your bare hands and then put on gloves. The inside of the tube should still be sterile, and so you can use it without worries. However, while pipetting, the end of your tip touches the outside of the tube. Throw out the tip and start over. You can't trust the tip to not introduce skin cells, skin oils, skin bacteria, or other junk into the reaction. Don't sabotage a $200 experiment for the sake of $0.07 of tips. See number 2, above.

You will find that, for different experiments, there may be additional steps you take to keep your samples clean. These suggestions are just to get you started.

ALIQUOTING

Anything that is sensitive to freeze-thaw cycles but needs to be stored in your freezer should be aliquoted. Aliquoting is the practice of separating solutions into smaller batches in separate tubes. The idea is that if you only need to use 5 μl of your 400 μl of solution, you don't want to thaw all 400 μl and lose activity each freeze-thaw cycle. You want to aliquot your solution and keep the activity of your materials the same for each experiment. Most labs commonly aliquot antibody solutions, but it is helpful to aliquot PCR and size marker solutions.

Aliquoting

For protein and DNA size markers (called "ladders"), protease inhibitors, and other one-use solutions, it is best to determine what amount you need to use per experiment and aliquot that amount. For instance, I found that I can use 8 μl of a certain protein ladder per sodium dodecylsulfate polyacrylamide gel electrophoresis and get great resolution, so I aliquot the ladder at 8 μl per tube. I also use a GTP solution for electrophysiology at 100 μl per experiment for my intracellular solution, so the GTP aliquot is 100 μl. Aliquoting one-use solutions in this way means that I don't need to compare aliquots to determine if there's enough solution for an experiment; I just grab one tube and go. Aliquoting, in this situation, saves the activity of your material and saves you time later.

I have aliquoted almost every primary antibody that I have come across. The main exceptions to this aliquoting rule have been antibodies shipped in glycerol or antibodies that are to be kept at 4°C. Antibodies in glycerol (or something with similar viscosity) shouldn't be aliquoted with additional molecular biology-grade glycerol unless you specifically want to dilute them further (always make sure that your glycerol or diluent is free of bacteria and molds). You can aliquot them and they don't appear to lose activity. Aliquoting secondary and tertiary antibodies may or may not be beneficial to you due to the sheer amount of aliquots you would generate; use your best judgment.

You can aliquot antibodies either in their stock solution or in a 50% glycerol solution (if you have 100 μl of antibody solution, just add 100 μl glycerol and aliquot); glycerol lowers the freezing point of the solution below −20°C. Typically, I follow the instructions of the manufacturer for how to aliquot the antibodies. I have aliquoted Chemicon and Invitrogen antibodies without glycerol (5 μl per tube) and left them at −20°C for years with no loss of activity or volume. Jackson ImmunoResearch recommends diluting their antibodies in water and glycerol, and I have followed those directions (they stay good for many years).

Typically, I put my aliquots in small microcentrifuge tubes and put those tubes into a 50 ml centrifuge tube. I label the small tubes with some identifier and the 50 ml tube with the antibody's information (catalog number, manufacturer, host species, target, date I aliquoted the antibody, my initials so people know who to blame, and the volume of aliquots) and store that tube at −20°C (−80°C if the manufacturer says to do so). Label all the tubes, unless you want to be caught in the middle of a hurried, crucial experiment with three unlabeled tubes and the rewards of your previous hubris. If your antibodies are conjugated to fluorescent markers, protect them from light (light exposure will decrease their activity over time while they refrigerate). If your antibodies are conjugated to horseradish peroxidase (HRP), or you plan to use them in chemiluminescent reactions, do not mix them with sodium azide (a preservative that inhibits HRP, among other things).

As for the volume selection, some people claim that aliquots less than 10 µl are more prone to evaporate; that hasn't been my experience. If using antibodies for multiple applications, then the aliquot volume is largely up to individual taste. Different antibody techniques will require different antibody concentrations, so you're not saving yourself any work by choosing different volumes. I haven't noticed a great loss of antibody activity with five freeze-thaw cycles, so I aliquot at 5 µl even if I only use 1 µl at a time.

Additionally, aliquoting can increase your quality of life in the lab. If you have a collaborator who wants to use some of your great antibody, you can give them an aliquot and not worry about whether they contaminated the whole tube or thawed it for too long—if they did any of that, it only ruined the 5 µl you gave them. If you are training someone new to the techniques, or have someone you don't trust to treat your materials correctly, you can give them an aliquot and put the rest in your own special place. Even if you don't hide the item, it's good for peace of mind that only a little of your investment is being affected at a time.

Long story short: aliquoting is a classic example of planning, where increased effort up front saves you effort later. While aliquoting can take up a day of work, the benefits are worth it. Plan your experiment and, where possible, plan your aliquoting.

CHAPTER 2

DNA and RNA Extraction Protocols

PHENOL:CHLOROFORM (TRIZOL™) EXTRACTION NOTES

I have used TRIzol™[1] (Invitrogen; similar stuff is sold by different companies as TRI Reagent™,[2] RNAzol™,[3] QIAzol™,[4] and other trade names, but they're all used in phenol:chloroform extraction of nucleic acids) and different kits to extract RNA. Phenol:chloroform extraction can be used to extract DNA, RNA (including short RNAs and microRNAs), and proteins, but it is only really useful for DNA and RNA. The phenol:chloroform extraction process works by denaturing proteins (including RNases, DNases, proteases, and any protein you want to examine) in order to preserve the nucleic acids—so extracting proteins in this way is a race against time to inactivate sample-destroying enzymes while preserving the proteins you want to examine. While the use of phenol:chloroform extraction appears to save posttranslational modifications (phosphatases appear to be inactivated over the course of this process, leaving phosphorylated proteins intact), the efficiency of total protein extraction is low. Roughly half of the protein samples I've collected using this method have such low concentrations that they couldn't be used at all for immunoblotting, while the other half were acceptable but minute compared to total protein extraction protocols described later in Chapter 6.

Samples stored in ice-cold or frozen guanidium isothiocyanate solution (TRIzol™, for example) do not remain viable; either of these storage methods result in denatured nucleic acids that have no use in any downstream applications. Store samples in RNAlater™[5] (or equivalent solutions available from a number of manufacturers) and follow the

[1]TRIzol® is a registered trademark of Molecular Research Center, Inc.
[2]TRI Reagent® is a registered trademark of Molecular Research Center, Inc.
[3]RNAzol® is a registered trademark of Molecular Research Center, Inc.
[4]QIAzol® is a trademark of QIAGEN Group.
[5]RNAlater® is a trademark of AMBION, Inc.

Basic Molecular Protocols in Neuroscience: Tips, Tricks, and Pitfalls. DOI: http://dx.doi.org/10.1016/B978-0-12-801461-5.00002-2

manufacturer's instructions or flash-freeze tissue samples. Samples stored in RNAlater™ and left at −20°C have remained viable for years. You can retrieve proteins from samples stored in RNAlater™ by whatever method you choose, though histology can be complicated by the method of fixation employed by RNAlater™. I prefer to perform histology with perfused tissue (see Chapter 9).

I typically follow phenol:chloroform extraction with a commercial kit's RNA cleanup protocol, followed by vacuum centrifugation to evaporate leftover phenols (which can interfere with downstream reactions) and to increase the concentration of nucleic acid in the sample. In my experience, phenol:chloroform extraction followed by a kit cleanup protocol (such as the RNEasy™[6] kit from Qiagen or the QuickRNA™[7] kit from Zymo Research) gives higher RNA purity than using phenol:chloroform extraction alone and higher RNA concentration than using the kit alone. This may be due to the high fat content of brain tissue; I think the guanidium isothiocyanate breaks fatty tissue down better than some kits, allowing for more RNA to be extracted from the tissue.

When you have extracted the RNA/DNA, I recommend using a spectrophotometer to measure RNA/DNA concentration and purity. The spectrophotometer uses three wavelengths of light to estimate these parameters: 280, 260, and 230 nm. Nucleotides (both DNA and RNA) respond to 260 nm light, while proteins respond to 280 nm light, and organic solvents respond to 230 nm light. As such, comparing the outputs of these wavelengths gives a measure of sample purity. The 260/280 ratio is a measure of protein contamination of your sample, while the 260/230 ratio is a measure of contamination by organic solvents that inhibit downstream experiments. The 260/280 ratio should be between 1.9 and 2.1 for RNA. The 260/230 ratio has come into focus lately, as some groups report that it should also be close to 2.0. In my experience, the numbers generated for the 260/230 ratio have not corresponded to any difference in my quantitative polymerase chain reaction (qPCR).

---

[6]RNEasy® is a trademark of QIAGEN Group.
[7]QuickRNA is a trademark of Zymo Research.

RNA method selection

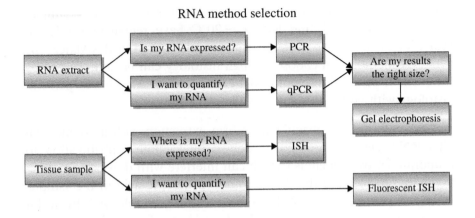

PHENOL:CHLOROFORM EXTRACTION PROTOCOL

1. Dissect the tissue from the animal and place in RNAlater™ according to manufacturer instructions. For cell culture, cells will need to be removed from culture and pelleted; store pellet in RNAlater™. Store samples according to manufacturer instructions.

   My experience has been that phenol:chloroform extraction of RNA is useless without RNAlater™ or its equivalent; sample quality is inconsistent without preserving the RNA in some fashion. Be quick with the dissection, as lost time is lost sample.

2. When you are ready to extract RNA, aliquot cold guanidium isothiocyanate solution (TRIzol™ or the like) into individual tubes and leave on ice. Volume is in accordance with manufacturer's instructions (1 ml per 100 mg tissue, generally).

   Don't put your samples in polyethylene tubes (chloroform melts polyethylene plastic); polypropylene tubes are fine. I find that commercially available guanidium isothiocyanate solutions are more than acceptable for RNA extraction. Making up your own solution from scratch is an unnecessary headache.

3. Immediately before homogenization, transfer tissue from RNAlater to guanidium solution.

   Brain tissue stored in RNAlater™ becomes tough and rubbery. Gently use cleaned forceps to transfer tissue. Don't spill any guanidium isothiocyanate solution on yourself; it burns the skin and leaves scars for years.

4. Homogenize sample using a sonicator or rotor-stator homogenizer, typically at 50% or less of total amplitude.

I have used sonicators, rotor-stator homogenizers, and hand-held homogenizers with disposable nuclease-free pestles. While the disposable pestles can work, their performance is highly variable (likely because it's harder to generate the necessary friction between two plastic surfaces), and the amount of RNA extracted is very little compared to a rotor-stator homogenizer, most likely due to incomplete tissue dissociation. I have used a rotor-stator homogenizer with separate autoclave-compatible pestles, which have kept quite clean and I have had no problems with RNase. I would not use disposable pestles again unless my life depended on it.

5. Allow homogenized samples to sit in guanidium solution at room temperature for 5 min.
6. Per 1 ml of guanidium solution, add 200 µl of chloroform. Vortex the sample to mix and let sit for 3 min at room temperature.

> *Vortexing is crucial; the clear chloroform will sink to the bottom after centrifugation and will not mix with the solution if you do not vortex the sample. If you lack a vortexer, you need something that will generate strong force to mix the sample.*

7. Centrifuge samples at 12,000g for 10 min at 4°C.
8. You should see a clear aqueous layer at the top, a milky white layer underneath the aqueous layer, and an organic layer (typically pink; depends on the manufacturer of your guanidium solution) at the bottom of the tube. Transfer the clear aqueous layer to a new microcentrifuge tube (used for RNA extraction; steps 9–16). For DNA and protein extractions, keep the remainder on ice (steps 17–28 for DNA, steps 29–37 for protein).

> *It is very important that the pipette tip does not disturb the milky white layer. Touching the white layer inevitably allows the introduction of the white and organic layers to the aqueous layer, which will add DNA and protein contamination to your RNA extract. As one gets more skilled with the technique, the amount of aqueous layer that one can extract without affecting the other layers increases; however, do not sacrifice your sample for that extra 50 µl. This step is the source of most, if not all, frustration associated with this technique.*

RNA EXTRACTION

9. To the clear aqueous layer, add 200 µl chloroform, vortex the sample to mix, and let sit for 3 min at room temperature.

There are two purposes to this additional step. The first is to increase the purity of the sample, which is especially important for quantitative PCR. The second purpose is to effectively idiot-proof the protocol by ensuring that accidentally touching the DNA layer does not ruin your subsequent extraction steps. As with the previous chloroform step, vortexing is crucial.

10. Centrifuge samples at 12,000g for 10 min at 4°C.
11. Transfer top aqueous layer to a new tube.

If the aqueous layer is not liquid but is instead foam, throw out your guanidium isothiocyanate solution. The pH is incorrect and you're better off just replacing the solution than to try to pH it properly and fix it. This seems to occur after 5 years or more of storage.

12. Add 500 μl of isopropanol to the aqueous layer and leave in −20°C overnight.

The typical protocol suggests that samples sit for 10 min in isopropanol and are then spun, but I have had greater success using this overnight incubation. This gives more time for the isopropanol to bind the RNA and gives greater flexibility in experimental planning; I have left samples in isopropanol for over a week and had no problems with purity or stability of the resulting RNA.

13. Retrieve samples from −20°C and centrifuge at 12,000g for 10 min at 4°C.
14. A pellet should be visible on the wall of the tube. Wash the pellet using 100% ethanol. Remove the excess liquid and invert to air dry; this typically takes 10−20 min.

I use a wash bottle and point the stream of ethanol at the wall and not directly on the pellet, so that the pellet isn't removed. The pellet is off-white and sometimes slightly yellowed in appearance. Wait until the samples are completely dried.

15. Resuspend pellet in RNase- and DNase-free water.

I typically use the volume of nuclease-free water recommended for the cleanup kit that I want to use; if you're just getting RNA from here without additional cleanup, use any amount you want. Some protocols suggest that you heat the samples to improve resuspension of the RNA; I rarely had success when heating up the sample, and so I omit that step and have never had a problem since.

16. Begin cleanup protocol using a kit or begin evaluating samples; this step varies, depending on the kit of choice. When finished, store at

$-20°C$. Evaluate samples using a spectrophotometer (to determine concentration and purity) and gel electrophoresis (to assess RNA structural integrity).

Some groups prefer to store their RNA in Tris—ethylenediaminetetraacetic acid (EDTA) (TE) buffer. I have stored RNA for 5 years at $-20°C$ with little to no degradation in RNase- and DNase-free water. I prefer water to TE to avoid adding extra EDTA to any subsequent reactions (EDTA will bind magnesium, which is necessary for PCR). I have a TE buffer recipe at the end of this chapter if you wish to use it.

DNA EXTRACTION

17. Remove any remaining aqueous phase from the DNA/protein extract.
18. Add 300 μl of 100% ethanol per 1 ml guanidium solution to the sample. Vortex to mix and let sit for 3 min at room temperature.
19. Centrifuge samples at 2000g for 5 min at 4°C.
20. Transfer the colored supernatant (protein extract) to a new tube and keep on ice.
21. Wash the pellet with 1 ml of a sodium citrate/ethanol solution (0.1 M sodium citrate, 10% ethanol, pH 8.5), mix by inverting the tube, and let sit for 30 min at room temperature.
22. Centrifuge samples at 2000g for 5 min at 4°C.
23. Remove supernatant and wash pellet with 1 ml of the sodium citrate/ethanol solution, mix by inverting the tube, and let sit for 30 min at room temperature.
24. Centrifuge samples at 2000g for 5 min at 4°C.
25. Remove supernatant. Add 1.5 ml of 75% ethanol to tube, mix by inverting the tube, and let sit for 10 min at room temperature.
26. Centrifuge samples at 2000g for 5 min at 4°C.
27. Remove supernatant. Invert tube and allow pellet to air dry. Air drying typically takes 10–20 min.
28. Resuspend DNA pellet in 8 M NaOH and store at $-20°C$. Evaluate samples using a spectrophotometer and gel electrophoresis.

As with RNA extraction, the volume of NaOH used to resuspend the DNA is up to you. Increased volumes result in more dilute DNA. Do not use a vacuum microcentrifuge for this sample, as decreasing the volume increases the concentration of both your DNA and NaOH, possibly harming your DNA. I highly recommend using a cleanup protocol from a kit so that you can get your DNA out of NaOH solution and into either water or TE.

PROTEIN EXTRACTION

29. Add 1.5 ml of isopropanol to the protein extract per 1 ml of guanidium solution used. Mix by vortexing and let sit for 10 min at room temperature.

    Transfer to a larger tube that can hold at least 3 ml. If not, you won't be able to use these volumes and you will likely lose some protein.

30. Centrifuge samples at 12,000g for 10 min at 4°C.
31. Remove supernatant and wash pellet in 2 ml of a guanidine hydrochloride solution (0.3 M guanidine hydrochloride in 95% ethanol). Mix by inverting the tube and let sit for 20 min at room temperature.
32. Centrifuge samples at 7500g for 5 min at 4°C.
33. Repeat steps 31 and 32 two more times.

    This is a total of 60 min of washing and 15 min of centrifugation.

34. Remove supernatant and add 2 ml of 100% ethanol. Mix by inverting the tube and let sit for 20 min at room temperature.

    At this point, the pellet(s) may not stick to the wall of the tube. Be careful to keep your pellet in the tube.

35. Centrifuge samples at 7500g for 5 min at 4°C.
36. Remove supernatant and let air dry for 20 min.
37. To resuspend the protein pellet, I make two solutions: a 1% sodium dodecyl sulfate (SDS) solution and a solution of 8 M urea. I add 100 μl of each solution to the pellet and mix. Store at −20°C. Evaluate concentration using a spectrophotometer using 280 nm light.

    I got the idea from a report that claimed this to be the best method.[1] Protein evaluation, using this method, cannot be done using the Bradford method, as both the concentration of urea (4 M) and SDS (0.5%) interfere with the reaction. For Bradford reaction-compatible samples, the urea must be omitted and the SDS concentration must be less than 0.1%. There are pink- and white-insoluble pellets surrounded by solution. The soluble protein is in the solution, not the pellets, and that is the protein you can use for immunoblotting or immunoprecipitation.

KIT EXTRACTION OF RNA NOTES

I have used kits from multiple manufacturers to extract RNA and DNA. The kits work well, though there are still a couple of important changes to remember (these are listed as "optional" steps in the kit protocols):

1. At the end of the protocols, when collecting RNA from the spin columns, always let the final eluent (water or buffer that the RNA will be collected in) sit in the column for 1 min before spinning; this increases the yield.
2. After the first RNA collection, put another volume of eluent in the column and let it sit for 1 min before spinning again. This second spin always has a significant amount of RNA in it, which I have verified using a spectrophotometer. I combine the two eluates into a single tube.
3. Spin samples (even briefly) in a vacuum centrifuge afterward to evaporate leftover phenols.
4. Include a DNase reaction to reduce or eliminate genomic DNA contamination of RNA samples.

I find that kits, used in conjunction with phenol:chloroform extraction, allow for excellent, closer-to-maximized sample extraction (there is always some loss) with a high degree of sample purity. Using a kit alone is feasible for most applications, and very useful when teaching the inexperienced (undergraduates, new technicians, and the like). While many kits cannot collect microRNAs or short RNAs as well as phenol:chloroform extraction, newer kits are (at the time of this writing) becoming available for these targets.

SOLUTION RECIPES

0.1M sodium citrate/10% ethanol, pH 8.5	
Na$_3$ citrate	5.16 g
100% ethanol	20 ml
dH$_2$O	To 200 ml

pH solution to 8.5.
Store in an airtight bottle at room temperature.
Keeps for a long time.

0.3M Guanidine HCl in 95% ethanol	
Guanidine HCl	5.73 g
95% ethanol	To 200 ml

8M NaOH	
Sodium hydroxide (NaOH)	3.199 g
dH$_2$O	To 10 ml

1% SDS	
Sodium dodecyl sulfate (SDS)	100 mg
dH$_2$O	To 10 ml

Alternatively, if you have 10% SDS solution from
immunoblotting, you could simply dilute that.
Wear eye, nose, and mouth protection.

8M Urea	
Urea	4.81 g
dH$_2$O	To 1 L

This is an endothermic solution, so it is slower to mix.
Higher concentrations of urea won't go into solution
completely. Do not heat solution!

Nuclease-free water (0.1% DEPC), aka RNase-, DNase-free water	
Diethylpyrocarbonate (DEPC)	1 ml
dH$_2$O	To 1 L

Mix DEPC into the water by agitation and let sit for at least 2 h (until globules disappear).
Autoclave.
Alternatively, nuclease-free water can be bought from many sources relatively cheaply. DEPC solutions should not
come into contact with Tris solutions before autoclaving, as Tris inactivates the DEPC. Autoclaving prevents DEPC
from inactivating any newly introduced RNases, so keep your stocks of RNase-free water clean.

Tris-EDTA (TE) buffer	
Tris base	0.607 g
EDTA	0.186 g
dH$_2$O	500 ml

pH to 8.0 using hydrochloric acid (HCl). Autoclave to sterilize.
Many groups make stocks of the Tris and EDTA solutions, if you find these amounts too small to weigh accurately.

CHAPTER 3

Agarose Gel Electrophoresis

ELECTROPHORESIS NOTES

We use electrophoresis to separate nucleic acids or proteins by size and/or charge, depending on what we've put into our solutions (more of an issue with SDS-PAGE; see Chapter 7). Basically, in gel electrophoresis, you put your samples into a gel matrix and your negatively charged nucleic acids (negatively charged due to their phosphate "backbone") are repelled from the negative pole of your electrodes and toward the positive pole at the other end of the gel. Larger RNAs and DNAs (and proteins, in SDS-PAGE) have a harder time moving through the gel matrix and therefore you can separate your RNA/DNA/proteins by size. Additionally, this is the only way to check your results from end-point PCR (see PCR section in Chapter 4). In my experience, running electrophoresis gels is a balance between efficiency and beauty: running the gel at higher voltages will move the samples along the gel faster, but the resulting bands are more likely to be smeared (and if you set the voltage high enough, you'll melt the gel and destroy all your hard work). Run the gels at lower voltages and you'll get a prettier picture, but you'll need to wait.

I have experience with three types of gel electrophoresis: agarose gel electrophoresis, denaturing (formaldehyde) gel electrophoresis, and sodium dodecylsulfate polyacrylamide gel electrophoresis (SDS-PAGE). Agarose gels are cheap and can be used to examine RNA or DNA stability. Denaturing gel electrophoresis can be used with RNA for northern blotting, but is mostly unnecessary if your goal is to examine RNA stability, which can be examined using agarose gel electrophoresis. While one can make RNase-free solutions for agarose gel electrophoresis, I only find it necessary for the loading buffer and water but not the gel or buffer. RNase-free solutions are necessary if one chooses to perform northern blots, which are not discussed in this book. SDS-PAGE is most useful for separating proteins and is therefore discussed in Chapter 7.

In agarose gel electrophoresis, one of two buffers is used: Tris-Acetate–EDTA (TAE) or Tris-Borate–EDTA (TBE). TBE has a higher buffering capacity than TAE. TBE buffer components

Basic Molecular Protocols in Neuroscience: Tips, Tricks, and Pitfalls. DOI: http://dx.doi.org/10.1016/B978-0-12-801461-5.00003-4

precipitate out of solution when stored at higher concentrations ($10\times$ solution, for example), so I keep a $0.5\times$ stock and avoid precipitation and stability issues.

Visualizing the results from agarose gel electrophoresis is typically performed one of two ways: including a DNA-binding fluorophore in the gel solution, or keeping that fluorophore in a buffer solution, washing the agarose gel in it, and using that solution for multiple gels. In both cases, you will use an ultraviolet light to excite the fluorophore and see the experiment's results. The most common DNA-binding fluorophore is ethidium bromide (EtBr). EtBr is toxic and your local environmental health and safety board has recommendations for how to handle it; do what they tell you to do. Another common fluorophore, and the one I use, is SYBR Safe™[1] (Invitrogen). SYBR Safe™ does better on toxicity tests in lab animals than EtBr, but consuming a DNA-binding anything isn't a good idea. In my experience, SYBR Safe™ degrades over time in free solution but not in its stored $10,000\times$ concentrate. As such, I get consistent results by including SYBR Safe™ (or EtBr) in my gel solution after it has cooled enough, before casting the gel. When the fluorophore is free in solution, one needs to incubate gels for longer periods of time to get equivalent staining in reused solution; however, if you're examining gels for fluorescence, then you can't tell if a lack of fluorescence is because the experiment didn't work, you didn't incubate in solution for long enough, or you need to make fresh fluorophore solution. Further, washing in fluorophore solution typically requires a wash in water to decrease the background noise, and you can leave it in water for so long that you wash out the fluorophore. In all, I think that including your fluorophore in the gel is the way to go, and you use less over time if you aren't replacing your staining solution every couple of days.

A good reference for this section is the book *At the Bench* by Kathy Barker.[2]

AGAROSE GEL ELECTROPHORESIS PROTOCOL

1. After identifying your desired casting apparatus (including the number of wells you want per gel), make an appropriate amount of agarose gel solution in some container (I prefer an Erlenmeyer flask).

[1]SYBR Safe® is a trademark of Molecular Probes, Inc.

I use 1.6% agarose gels for separating RNA or PCR results. Increasing the amount of agarose will slow passage through the gel, allowing you to visualize smaller strands of nucleic acid that would normally run off the gel. Decreasing the amount of agarose will speed passage, allowing smaller strands to run off the gel but increasing the likelihood that you'll resolve larger bands appropriately. The amount of solution you make and how the gel is poured depend on the apparatus you use for your gels; follow the manufacturer's advice.

2. When the gel solution cools enough to allow you to touch the container without burning your hand, add the DNA-binding fluorophore. Shake to mix and pour into gel caster.

   Use the back of your hand to test the temperature. If at any point you want to draw your hand back, do so and continue to wait. You need it cool enough that you can handle the container and not destroy the fluorophore, but hot enough that your solution isn't solidifying.

3. Using a pipette tip (you can use the tip you used for the fluorophore), move any air bubbles in your gel to the outer edges and place your well caster at the appropriate spot. Remember that your samples will travel from the anode (−) to the cathode (+), so don't put your wells by the cathode.

   Any air bubbles in the gel will stop the nucleic acids that meet it, and so you will see a lot of fluorescence at the air bubble site that doesn't move further. This obviously defeats the purpose of the experiment, so move the air bubbles and get them out of the way of your samples.

4. Allow gel to cool and solidify; the time required is dependent on the size of your gel.

   The gel should turn an opaque white. If you're not sure, gently shake the container and see if you can detect any liquid moving (if so, leave it alone to finish solidifying). Now that the gel is setting, wash your agarose solution flask immediately before the liquid solidifies; solid agarose is most easily removed by melting it.

5. Remove well casters and gel casters and cover gel in the buffer you used to make the gel (for me, 0.5× TBE). Make sure that a layer of liquid covers the top of the gel or the circuit will be broken and your nucleic acids won't travel.

   Basically, keep the circuit complete but don't put in so much buffer that you begin to spill it everywhere.

6. Mix samples with loading buffer. Keep original samples (and possibly mixed buffer/sample) on ice.

   The loading buffer serves two purposes: it increases sample density (thanks to the glycerol) so that the samples load into the wells more easily (otherwise they'd diffuse in the surrounding buffer) and you can keep track of where your samples are in the gel (thanks to the dyes). I use 6× loading buffer, which you want at a working concentration of 1×. For example, I might use 2 µl of sample, 3 µl of water, and 1 µl of loading buffer. I typically mix buffer and water on a small sheet of waxy paper (Parafilm™[2]), then mix with sample and load into the gel immediately.

7. Load samples in the gel, including a DNA "ladder" of appropriate size.

   The appropriate size marker ("ladder") is chosen by the size of nucleic acid strands that you're examining: if your target is 1000 bases long, you want a ladder that includes a band at 1000. I don't want to make my own ladder; companies make lots of the stuff and they make it cheaper than we could at home. Buy some ladders and read the manufacturer's specifications about what size of bands their ladders generate. Most ladders come in loading buffer, and so you can insert them into a well straight from the bottle, though it can help to aliquot the ladder at the volume you typically use (see the Aliquoting section in Chapter 1).

8. Set the voltage for running the gel. Check amperage to make sure it doesn't fluctuate.

   This is where the balance between efficiency and beauty comes into play. I use a minigel system and I run my gels at 72 V for 90−120 min. This generates beautiful pictures that show good resolution of my bands, little to no smearing (unless that's the fault of the sample), and both large and small bands (25−1000 bp). Because I only want to run a given gel once, I wait for the 90 min and make sure that all my gels are "figure ready" so that I don't need to rerun my experiments in order to get a prettier picture.

9. When the gel is done, visualize the bands using UV light. This can be done on a gel imaging system or a UV plate.

   For DNA or RNA gel electrophoresis, your options really are smears, crisp bands, or blank lanes. Similarly to immunoblotting, your data will be colloquially called "bands."

---

[2]Parafilm® is a registered trademark of Bemis Company, Inc.

10. Take pictures for posterity.

If you don't take pictures when you run your gel, you are likely to need the results later and possibly have to rerun the electrophoresis to get the information. Taking a picture that you can label and examine later is of immense value and can be safely tucked into a lab notebook that you can check anytime. Make sure that when you want to run electrophoresis that you have some way to take pictures and record your data.

SOLUTION RECIPES

0.5× TBE		
Tris base	108 g	*To make a 10× solution, add water to 1 l instead of 20 l. There*
Boric acid	55 g	*are claims that 5× solution is more stable than 10×; to make a*
Na$_2$ EDTA	7.5 g	*5× solution, add water to 5 l instead of 20 l. I have never had*
dH$_2$O	To 20 L	*stability or precipitation issues with the 0.5× solution in 7 years.*

6× Gel loading buffer	
Glycerol	15 ml
Bromophenol blue	0.25 g
Xylene cyanol	0.25 g
dH$_2$O	To 100 ml
Aliquot, store at −20°C. Can keep an aliquot at 4°C with no ill-effects.	

1.6% Agarose gel	
Agarose	0.8 g
0.5 × TBE	50 ml
SYBR Safe™ (10,000×)	5 µl
Add SYBR when container is cool enough to touch.	

CHAPTER 4

Reverse Transcription (RT) and Polymerase Chain Reaction (PCR)

REVERSE TRANSCRIPTION NOTES

Reverse transcription (RT) is the process by which RNA is converted into complementary DNA (cDNA) by the retroviral enzyme reverse transcriptase. A retrovirus is a virus capable of putting its RNA genome into a host cell via RT (in order to reproduce) and is thus the source of the reverse transcriptase we use. All retroviruses are nasty, but the one that gets the most press is the human immunodeficiency virus; this gives you an idea of how dangerous they can be.

We perform RT because DNA is much more stable than RNA and can undergo PCR. While RNA will denature if you breathe in its general direction, cDNA will remain much more stable (I've kept some cDNA at $-20°C$ for over 4 years and it still works like the day I prepared it). There are a number of kits for RT, both one-step RT (combined with PCR) and two-step RT (separate from PCR). I find the kits easy to use and I certainly don't want to spend time with any retroviruses to get my own stock of reverse transcriptase. I prefer two-step RT reactions for four reasons: two-step RT is more sensitive than one-step RT, I have greater control over the reaction, I can judge the efficiency of the RT separately from the PCR reaction's efficiency, and two-step RT allows me to generate a large amount of cDNA from a single reaction that I can spread across multiple tests, whereas a one-step RT reaction means that my RNA sample is now a limiting factor. I always dilute my cDNAs; more on that later.

I have tested kits from multiple vendors against each other. There are four factors that influence which RT kit I purchase:

1. *How much RNA can I put into the reaction?* One RT kit allows up to 5 µg of total RNA in the reaction, whereas others allow a maximum of 1 µg. More input RNA at the beginning allows me to see low-abundance RNAs more easily via PCR.

Basic Molecular Protocols in Neuroscience: Tips, Tricks, and Pitfalls. DOI: http://dx.doi.org/10.1016/B978-0-12-801461-5.00004-6

2. *Which kit has the most efficient RT reaction?* When I tested the results of two kits against each other, one of the kits generally gave me less cDNA (as evidenced by higher C_Ts in quantitative PCR; see qPCR section for an explanation of C_Ts).
3. *Which kit is most consistent?* When I tested two putative reference genes, β-actin and glyceraldehyde-3-phosphate dehydrogenase (GAPDH), for the same samples between two RT kits, β-actin showed a $10 - C_T$ difference within my samples generated using the first kit, but those same samples, when reacted with the second kit, had both lower C_T values (more cDNA) and showed no difference between each other.
4. *Which kit allows the most flexibility?* Can I use both oligo$_{dT(20)}$ and random hexamers in the same reaction at concentrations that I determine? I add both random hexamers and oligo$_{dT(20)}$ to the reaction to prime it from both ends, following the suggestions of Resuehr and Speiss[3] and Ståhlberg et al.[4]

In general, I follow the manufacturer's instructions for the kit. I don't bother priming my RT reaction with gene-specific primers because the efficiency of the RT is much lower using primers than using the oligo/hexamer combination I use or oligo$_{dT}$ or hexamers alone. Finally, I dilute my cDNAs with nuclease-free water after the reaction, as described by Resuehr and Speiss (2003). This accomplishes two things: first, you can use your cDNA for more reactions than if you left it undiluted; second, some components of the RT reaction can inhibit downstream PCR, and diluting the sample dilutes those inhibitors as well (it also dilutes the stabilizing buffer, but the buffer still works). While Resuehr and Speiss reported using a 1:20 dilution of the cDNA in nuclease-free water, I use a 1:10 dilution because I am examining low-abundance messenger RNAs (mRNAs). For most experiments, you are throwing money down the drain if you don't dilute your cDNA, because PCR will still detect mRNAs in thoroughly dilute samples and amplify it over 40 cycles (see Polymerase Chain Reaction (PCR) Notes section).

POLYMERASE CHAIN REACTION (PCR) NOTES

PCR stands for Polymerase Chain Reaction. PCR works by doubling the number of copies of your DNA (or cDNA) of interest, by using the bare minimum that a cell would require to replicate its DNA. PCR is generally credited with creating the molecular biology revolution in the 1980s and beyond. Anyone reading this should know that most of the molecular biology revolution was due to the discovery of *Thermus aquaticus*, a small bacterium discovered in deep-sea thermal vents. Initially, researchers had to replace their polymerase with every round of PCR, because polymerases from other bacteria (like *Escherichia coli*) would break down at 95°C. *T. aquaticus* changed all that because the *T. aquaticus* polymerase wouldn't break down during the PCR process. You're allowed to automate your PCR because of a relatively obscure bacterium. This is why most of the polymerases you can buy for PCR have the name "Taq" in there somewhere; it's shorthand for *T. aquaticus*, which has the original polymerase that's been repeatedly bioengineered for as much efficiency as possible.

PCR allows us to double the number of DNA strands of interest per round of the reaction, which then allows us to see even small changes in picograms (that's 10^{-12} grams) of DNA or cDNA; that's amazing. However, the amazing part of it is also its downfall: the sensitivity of PCR can lead to lots of false positives, which means that you need to keep positive controls, negative controls, and other controls in mind for your experiment. For instance, I use a primer set that spans exon junctions to test for genomic DNA contamination of my cDNA sample; only genomic DNA should have the introns, and so will give me a larger product than mRNA converted into cDNA. This is but one of the controls you can (and should) use.

While there are many variants of PCR, my main experience is with traditional end-point PCR, reverse-transcription PCR (RT-PCR), and quantitative PCR (qPCR). In the literature, qPCR is also called real-time PCR, and also abbreviated RT-PCR in the literature, in case you wanted some confusion. I only call it qPCR and will admonish you to do the same to avoid confusion. Reverse-transcription quantitative PCR (RT-qPCR or qRT-PCR) also exists but has the same issues that RT-PCR has (see RT-PCR section). qPCR has some additional caveats that PCR and RT-PCR do not and thus has its own section. There is also single-cell PCR, which is like any other kind of PCR except that you must get DNA or RNA from a single cell, which means you must work with a drastically lower amount of DNA or RNA. The latest development as of this writing is digital PCR and digital droplet PCR, which are much more quantitative than qPCR at the cost of much greater expense to run the reactions. Single-cell and digital PCR protocols are not covered in this book. While one could also use northern blotting to detect RNA instead of RT-PCR, northern blotting has additional caveats, is not as sensitive as PCR, and is not in widespread use anymore for detecting RNA (though there are still some uses for it). As such, northern blotting is also not discussed in this book.

END-POINT PCR

Traditional PCR is sometimes called end-point PCR because there's not much else you can do with it but check the results on an electrophoresis gel and see what you made, or the product gets used in another application entirely (making transgenic animals or injecting transcripts into cells, among other things). Data analysis of your PCR results involves gel electrophoresis—check if the products are there and at the right weights. While you can make your own PCR reaction buffer, that would require you to have your own supply of polymerase. Since I did not, I simply ordered my polymerases from different companies. Different polymerases have different levels of efficiency, and you should test each new polymerase before you begin your experiments in earnest.

I'll put a typical reaction setup in a table, from a purchased PCR kit:

PCR Reaction Mix	
PCR reaction buffer	*Buffers the reaction*
Mg^{2+}	*With water, stabilizes the DNA's structure*
dNTPs	*Provides bases for new DNA replicates*
Primers	*Guide polymerase to your site of interest*
Sample	*No sample, no experiment*
Polymerase	*No polymerase, no experiment*
Water	*Ensures consistent volumes*

If you buy a commercial polymerase, it should come with everything except primers and sample (most companies include nuclease-free water in the box). Determining the proper concentration of Mg^{2+} is for the advanced practitioners of the technique; it's not something that the typical PCR user ever has to adjust from what you get in a kit or a commercial polymerase mix. If you think your reaction is expiring before it should, you can add more Mg^{2+} to stabilize your DNA and primers. Primer sets can be found in publications and online databases, depending on your model species, and most common targets have primer sets available online that have been verified by sequencing or some other means (but see Primer Design section).

PCR Cycles (30−40 Cycles of Denaturing, Annealing, and Extension)	
Hot-start step	*Knocks inhibitor off Taq*
Denaturing step	*Separate DNA strands*
Annealing step	*Allow primer binding*
Extension step	*Allows polymerase binding*
Final extension	*Finish extension*
Holding	*Keeps product intact*

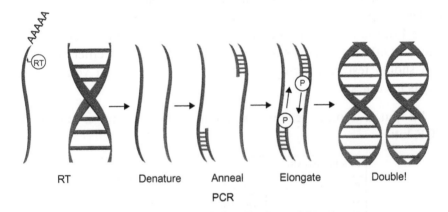

RT   Denature   Anneal   Elongate   Double!

PCR

As for the PCR steps, all steps (except the annealing step) are dictated by the polymerase that you've purchased, determining both the temperature and the time required for each step. The hot-start step is only for polymerases that require it; hot-start means that the polymerase was bound to something that inactivates the enzyme until you apply heat and therefore are using it. Hot-start polymerases usually ship better and arrive functionally sound.

Annealing temperatures vary between 50°C and 65°C, and I find that it is best to test the proper annealing temperature for your primers when using your particular polymerase. Ideally, one is supposed to run a reaction with an annealing step 5°C below the melting temperature of the primers, but this varies. The issue with the annealing step is, as with so many techniques, balance: lower temperatures encourage primer binding, including nonspecific binding, while higher temperatures increase specificity but can prevent your primers from binding their intended target. You can also use touchdown PCR, a type of PCR where you vary the annealing temperature between cycles, to bypass this testing of annealing temperatures, but that requires machines capable of performing the touchdown procedure. I keep spare cDNA around so that I can test the annealing temperatures and efficiencies of newly designed primers before I get to the experiment.

Follow the instructions of the polymerase supplier for running cycles and reaction components, unless you have deduced that you need to change them. You can run between 30 and 40 cycles of denaturing, annealing, and extension steps in a PCR reaction; fewer cycles are less likely to reach the plateau point (where quantitative comparisons become impossible), but additional cycles increase the likelihood that you'll see something, including false positives.

PRIMER DESIGN

If you need to design primers yourself, there are multiple programs that can help you design primers: oligo3 is available via the NCBI web site; OligoExplorer[1] is freeware available online; different companies provide automated primer design software. I get the sequence of my target via NCBI's database, load it into OligoExplorer, and design primers that lack hairpins, self-annealing structures, and primer dimers (primers that bind to each other), as these show up as false positives. Again, if someone else has designed the primer already and verified it (very important: make sure they verified it, or you can spend your time realizing you've been duped), don't recreate the wheel; just use the primer set previously published. However, I have had the experience of using a published, "verified" primer set, and it turned to junk in my hands. If this happens, design a new set

---

[1]OligoExplorer™ is the property of GeneLink, Inc.

(or better yet, three sets) and order them because oligonucleotide primers are cheap.

When you design your primers, you will need both sense and antisense primers. "Sense" primers run 5′ to 3′ and are also called "forward" primers. "Antisense" primers are the opposite strand (3′−5′) and are also called "reverse" primers. Primers, ideally, have melting temperatures between 60°C and 65°C, are between 18−30 bp, and have a GC content between 35% and 65%, do not bind to each other or to themselves, and have a product size that works with your application. Basically, end-point (and one-step RT-) PCR primers have products between 200 and 1000 bp, and if you go longer than that, you're entering the specialized valley of long PCR, which will not be discussed in this volume. qPCR primers should have products between 75 and 300 bp.

I have used primers at lesser and greater concentrations than 500 nM per reaction and found that 500 nM primers (forward and reverse, aka sense and antisense) is the sweet spot for my experiments; lower than that, I can't see product, higher than that, I'm using more than I need or getting more primer dimers. I typically mix forward and reverse primers two dilution steps away from the original stocks, because I don't want to risk contaminating the original (and first dilution) stocks, and having the two in one tube can help save time. I also enjoy this setup because I can dilute my primers enough that they are at the correct concentration and I don't have to add additional water to my reaction (every additional pipetting step is another chance for contamination). For example, I keep primer stocks at 1.25 μM because 8 μl of 1.25 μM primers in a 20 μl PCR reaction yields a primer concentration of 500 nM.

If I am running my primers in a qPCR reaction, I test the efficiency of the primer set. Efficiencies are calculated by running a five-point (minimum) serial dilution series of a single cDNA sample through qPCR. I calculate the slope of that dilution curve (a downward, negative slope) and use the following calculation: $10^{(-1/\text{slope})} = $ reaction efficiency, where $2 = 100\%$ efficiency (a doubling per cycle) and $1 = 0\%$ efficiency (no change). I run these in triplicate and average the calculated efficiencies. Knowing the efficiency of the reaction allows me to calculate the true difference between my experimental and control reactions (see qPCR section).

REVERSE-TRANSCRIPTION PCR (RT-PCR)

RT-PCR is also called one-step RT, because there is only one tube to fill and therefore only one step between you and your data (for you, not the thermocycler). Most of the steps and procedure are exactly the same as with end-point PCR, except for the RT step at the beginning; you still follow the manufacturer's instructions for most of it and use the same annealing temperature that you empirically deduced (you did that, right?). I think RT-PCR is fine for a qualitative answer: you don't care how efficient the RT reaction is, just that it's efficient enough (one-step RT is typically less efficient than two-step RT). If you want a quantitative answer, you need to perform qPCR or a shiny new variant of qPCR. It is important to keep limitations in mind as you design your experiment.

QUANTITATIVE PCR (REAL-TIME PCR OR QPCR)

qPCR allows one to easily quantify the amount of mRNA present in the sample, either by absolute measurements or relative measurements (Integrated DNA Technologies provides a very nice guide,[5] and a necessary reference for this section is the MIQE guidelines[6]; read the MIQE guidelines). qPCR works by dyes that fluoresce when they bind double-stranded DNA (intercalating dyes; SYBR Green™,[2] for instance) or primers with fluorophores attached that fluoresce after both binding to the target DNA and interacting with each other (TaqMan™[3] probes, among the many new iterations available). TaqMan™ probes are more specific than intercalating dyes and allow for multiplexing (multiple reactions in one tube). Multiplexing is possible when using TaqMan™ probes (or similar products from various competitors) because you can get multiple probes that emit light at different wavelengths, which is not possible with SYBR Green™. The trade-off for multiplexing capability is that SYBR Green™ is comparatively cheap. Often, SYBR Green™ comes with ROX™,[4] which is a background fluorophore. The computer records the fluorescence of both SYBR Green™ and ROX™, and

---

[2]SYBR Green® is a registered trademark of Molecular Probes, Inc.
[3]TaqMan® is a registered trademark of Roche Molecular Systems, Inc.
[4]ROX is a trademark of Life Technologies Corporation.

fluctuations in ROX™ indicate evaporation of the reaction liquids or variation in your pipetting; in any case, this is handled by the computer and generally not handled by you. Some machines don't use ROX™; they use fluorescein as the background fluorophore instead. I haven't observed any real difference between ROX™ and fluorescein as far as efficiency goes.

Whatever you select as a fluorophore, keep it protected from light. Aliquot SYBR Green™ mix and keep it out of the light. If SYBR Green™ is kept in clear plastic and exposed to light, the fluorophore will lose activity and your qPCR will appear to plateau early and have lower fluorescence overall, even though the rest of the reaction is proceeding as normal. See PCR troubleshooting section for further details.

For the most part, qPCR occurs (for the user) in the same manner at traditional PCR: mix a number of chemicals together, put them in a thermocycler, and come back for data. There are a couple of key differences, however. qPCR products need to be (depending on which reference you're looking at) between 50 and 400 bp. I typically design between 75 and 200 bp, though one 248 bp primer set works perfectly fine. The reason for this is that you want to give enough time for the polymerase to finish the extension, because if it doesn't, your quantitation will be off from the true value. The analysis of qPCR data is also different—while you can run your reactions in a gel, they are analyzed on the computer. Your reaction is measured by the amount of light it generates. In the case of SYBR Green™, the more DNA that gets amplified, more SYBR Green™ binds, and more SYBR Green™ molecules fluoresce. The key number is the cycle at which your reaction crossed the threshold for light emission above background (set by the program or by the user)—the threshold cycle or C_T. Most qPCR analyses use the C_T value. You can manually adjust where the threshold is for your data and therefore adjust the C_Ts, but it's not often necessary (most programs these days do a good job, and adjustment is rare). If you do need to adjust the threshold, you need to choose the point where the linear phase begins and avoid choosing a threshold that includes background noise. You also cannot analyze the plateau, because you've reached the limit of detection and there is no quantitative data there. I prefer to leave the thresholds alone, in general, so that I avoid introducing bias into my data.

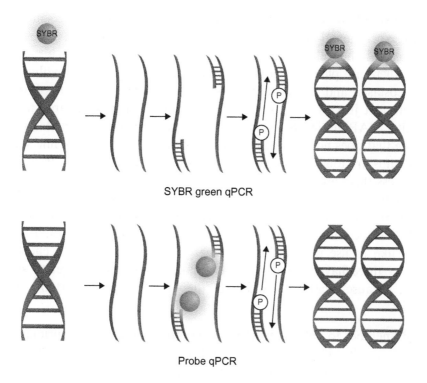

SYBR green qPCR

Probe qPCR

The fluorescence involved in qPCR makes another control possible: the melt curve. The thermocycler slowly raises the temperature and measures the fluorescence, which goes away when the transcript melts (denatures). Most primer design programs will suggest the melting temperature of the product along with information on your primers. Since you have an idea of the product melting temperature, you can check if the melting temperature of the actual product of your reaction was the same, or close to, the predicted melting temperature.

QUANTITATION METHODS

There are two main ways to quantify your qPCR data: absolute quantitation and relative quantitation. Absolute quantitation requires standard curves, on the same plate as your experiment, with known amounts of your target; there is variation in an identical reaction within a plate and between plates, so your standard curve must be present on each plate. You simply compare the experimental C_T values to the standard curve. Relative quantitation is between your samples, their experimental controls, and a reference gene. Reference genes are also called "housekeeping genes" in the literature, but this is a misnomer; there are plenty of good reference genes that are not involved in homeostasis, and homeostatic genes can change as a result of your experiment just as well as your targets (some experiments are known to alter GAPDH levels, everyone's favorite control). A reference gene is one that does not change as a result of your experiment. While some programs are touted to verify your reference genes, I haven't been able to access them and so I don't know if they work nor how they work. I prefer to test a panel of reference genes from all over the genome and compare them to each other; for example, you could use reference mRNAs encoding histone, ribosomal, and metabolic proteins. I typically like to use three reference genes in an experiment and check if two or three have similar patterns for the data; if not, I get to start over, because I can't tell which of the three (if any) report the true values.

Remember to calculate the reaction efficiencies for each primer set used in qPCR. This should be done for each reaction mix you test, as different polymerases and chemical compositions can alter your efficiency. Knowing the efficiency of the reaction allows you to calculate the real relative expression of your target.

RUNNING A PLATE AND DATA ANALYSIS

When I run an experiment, I run each sample in triplicate, so that any intraplate variation can be accounted for or at least examined. I chose triplicate rather than duplicate because, if my duplicates are different, how do I know which is correct? Triplicates at least have a chance for two of three to show a similar result. However, correct pipetting will generally give three close-to-identical reactions. I average my reactions and use that average value, omitting reactions that are two standard deviations away from my average (to control for outliers).

When I want to compare reactions across multiple plates, I include a "calibrator" sample (control) that is common to each plate, so that I can compare the samples to the calibrator and get an idea of how each sample changed. Since I calculated my efficiencies previously, I can use the method first described by Pfaffl[7] to quantify my real change in mRNA:

$$\text{Fold change:} \frac{\text{Efficiency (target)}^{(\text{average CT control}-\text{average CT sample})}}{\text{Efficiency (reference)}^{(\text{average CT control}-\text{average CT sample})}}$$

PCR CONTROLS

There are multiple controls that are good to run for the different kinds of PCR experiments that one might perform. Ideally, you run all the controls you can to make your story much more convincing, though some controls are prohibitively expensive. Use your best judgment. Additionally, always perform PCR reactions with replicates (I prefer to run in triplicate), as some PCR reactions may not run for different reasons, or you may identify differences between replicates.

1. *Water control (aka no template control, NTC)*: In this control, you put nuclease-free water in place of your cDNA template for your PCR reaction. This can identify if your primers are forming dimers and giving false positives for your PCR reaction, or possibly identify contamination in one or more of your reaction components (water, buffer, dNTPs, etc.).
2. *Negative control*: if you are examining a specific tissue or cell type for the presence of an mRNA, one of the easiest and best controls is to also examine tissue or a cell type that is known to not express that gene. If you get a band in your negative control, then your primers are not specific (or forming primer dimers, but you're redesigning the experiment in either case). An offshoot of this control (for qPCR) is to examine a tissue that does not respond to your experiment; if there is a change, that change should be due to something other than the experiment.
3. *Positive control*: if you are examining for the presence of a gene, this is a good control to include next to your negative control. A positive control can identify if you have inhibitors in your reaction (possibly from the RT or from your materials). As with negative controls, you can also examine targets that should respond to your experiment (such as making sure to examine a gene known to increase due to your treatment, for example).
4. *Intron control*: if you are examining mRNAs (and therefore performed a RT reaction first), then you are concerned about DNA

contamination (and therefore false positives). Therefore, it can be useful to design primers that span exon–exon junctions. If the amplicon is contaminating DNA, the product will be longer than if the amplicon came from RNA (cDNA), because the DNA contains introns and the cDNA should not. Empirically test any exon–exon primers to ensure they will amplify both cDNA and genomic DNA.

5. *No RT control*: if you do not want to design exon junction-spanning primers, you could also have a sample of RNA that did not undergo RT. Including that in your PCR would identify DNA contamination. I prefer exon-spanning primers, as that means I don't need to subject every sample to testing as a no RT control (but one primer set does double duty collecting data and testing for DNA contamination). Make sure to empirically test any exon-spanning primers with genomic DNA to make sure that they do amplify genomic DNA differently than cDNA.

6. *Multiplexing controls*: if you get into the wide world of multiplexing, there are a couple of considerations to keep in mind. First, you will need to run each primer/probe set individually, to understand how the results should appear. Then you will need to make sure that your different primers and probes do not interact with each other, or else you'll get erroneous data from primer dimers between different primer sets.

7. *No amplification control*: this control omits the polymerase from the reaction. This is more useful for qPCR, as it tests background fluorescence and probe stability in the absence of actual replication.

8. *Reference genes*: in qPCR, it is advantageous to test at least three reference genes to normalize data and ensure that the patterns are the same. If you only test one reference gene, then your data may be due to that reference gene fluctuating over the course of your experiment. If you test two reference genes and they give you conflicting results, you don't know which results to trust. Testing three allows you to see if two of three genes give the same (or similar) results.

9. *Sequencing*: if your product has not been sequenced before, you need to submit it for DNA sequencing to determine if your product is what you were looking for. If the primer set you are using has been previously published, check if they sequenced the results. If you get irregularities with your primer sets, sequencing can tell you relatively quickly if you need to redesign primers. This should be performed for any primer sets (at least once) that you want to publish!

PCR TROUBLESHOOTING

Issue	Possible Cause	Possible Solution
No amplification:	You omitted a reaction component, or a component was too dilute.	*Retry reaction.*
	Contamination in a component.	*Retry reaction with new materials.*
	You left your SYBR Green mix in the light; you did amplify, but there was no fluorescence for the machine to "see."	*Run on a gel to see if amplification occurred.*
	You did not optimize annealing temperatures.	*Back to the beginning.*
	Your primers are not specific.	*Redesign primers.*
	Your primers are degraded.	*Check primers in a reaction known to work.*
	Not enough template in the reaction.	*Increase amount of cDNA in the reaction.*
	Target below detection threshold.	*Increase amount of cDNA in the reaction. Test efficiency of primers.*
	RNA samples are no good.	*Collect new samples.*
Low signal:	Primers not completely resuspended.	*Resuspend primers by pipetting. Counting a certain number of "passes" can help.*
	Inhibitors in the reaction.	*Dilute template.*
	You used multiple sample prep methods in the same experiment.	*Some sample prep methods can alter the way the data looks. Use one method for the experiment.*
	C_T value higher than expected.	*Check primer efficiency and concentration.*
C_T lower than 15:	Too much template.	*Decrease amount of template.*
Noisy data:	ROX degraded or was low.	*Use fresh master mix.*
	Genomic DNA contamination.	*Include DNase in your RNA extraction.*
	Multiple products in one reaction.	*Redesign primers; primers aren't specific enough.*
Replicates were inconsistent due to pipetting errors.		*Increase number of passes to ensure proper mixing.*

CHAPTER 5

Antibodies and Titrations

Antibodies are glycoproteins (proteins that have oligosaccharide chains) that bind specifically to single antigens (ideally). Antigens can be protein sequences, physical shapes of proteins, or even chemical groups (when antibodies are made against chemicals like serotonin). However, an antibody may bind to any one or combination of factors regarding that antigen. A given antibody may bind amino acids in a specific sequence, the antigen's three-dimensional (3D) structure, a marker that indicates the species the antigen came from, and more. There are two main classes of antibodies that you can buy: polyclonal and monoclonal. There are also antibody subtypes, but subtypes other than immunoglobulin G (IgG) are mostly useful for multiplexing reactions that use primary antibodies from a single species. Immunoglobulin subtypes aren't used by most researchers because most commercial antibodies are IgG and most secondary antibodies bind IgG. Make sure that any secondary or tertiary antibodies you purchase bind your chosen immunoglobulin.

We get antibodies from a host animal, and polyclonal antibodies are taken directly from the blood of that host animal (called a bleed; bleeds are numbered by the number of times blood has been collected from the animal, so a fourth bleed is the fourth collection). Polyclonal antibodies (poly-clonal = many clones) are mixtures of all the antibodies that bound to the antigen, whether binding the 3D shape, the sequence, or any other marker that identified the antigen to the animal's immune system. As such, polyclonal antibodies can vary between host species, host animals, and lots, so you need to retest every new lot to make sure it works the same as the last lot. Polyclonal antibodies are typically affinity purified, which means that the antibody mixture (as opposed to the host animal) is exposed to the antigen, and only antibodies that bound the antigen are retained and sent to you. Antibodies that haven't been affinity purified still have antibodies against other antigens present (basically, whichever antibodies were in the host's bloodstream at the time of collection).

Monoclonal antibodies are single antibodies made by fusing antibody-producing cells with an immortalized cell line, resulting in a cell line

Basic Molecular Protocols in Neuroscience: Tips, Tricks, and Pitfalls. DOI: http://dx.doi.org/10.1016/B978-0-12-801461-5.00005-8

called a hybridoma. Hybridomas are separated by the individual antibodies that the different cells make, until you have a "pure" cell line that produces a single antibody. Because these cell lines are "immortal," you can make the same exact antibody (mono-clonal = one clone) for years and years, and that means that you shouldn't need to repeatedly test every new lot (a couple of papers have come out recently that suggest that, after a number of years, you should retest monoclonal antibodies due to mutations in the cell lines). Historically, monoclonal antibodies were made only with mouse cells, but mice don't have the best immune responses. New technology has made rabbit monoclonal antibodies possible, and rabbits have much better immune responses than mice. I have had great experiences with rabbit monoclonal antibodies.

For the purposes of this book, a "low" titer is more dilute (e.g., 1:100,000) than a "high" titer (e.g., 1:50). Antibodies are used in multiple applications, and in this book, they are described for use in immunoblotting, immunoprecipitation, and immunohistochemistry. Secondary antibodies can also be used in *in situ* hybridization but aren't necessary.

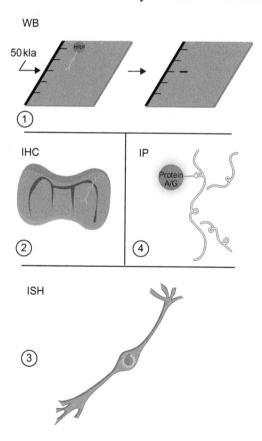

Antibody techniques refer to antibodies as primary, secondary, and tertiary (1°, 2°, and 3°, respectively). The primary antibody is the one that binds to your target. If you are examining β-amyloid protein and your antibody comes from a rabbit, then your primary antibody is called rabbit anti-β-amyloid. The secondary antibody is the one that binds to your primary antibody. Continuing the example, a goat anti-rabbit antibody would bind to the rabbit anti-β-amyloid, so the rabbit antibody is the primary and the goat antibody is the secondary. A tertiary antibody is an antibody that binds to your secondary antibody. Use of secondary and tertiary antibodies is a form of signal amplification. While you can conjugate your signal molecule to the primary antibody, this doesn't generate much signal for you to examine. Many secondary antibodies can bind the primary antibody (amplifying the signal), and many tertiary antibodies can bind each secondary antibody (further amplifying the signal). As such, this can rapidly amplify your signal so that you can examine it, quantify it, and take nice pictures.

Your choice of antibody can also influence what you use for your blocking step to block nonspecific interactions between antibodies and your samples. In general, you should use a blocker that will not give false positives from one or more of your antibodies. For example, if your primary antibody is from rabbit and your secondary antibody is from goat, then blocking with goat serum will result in low background (the goat isn't making antibodies against itself; an organism making antibodies against its own proteins is called autoimmune disease). If you were to block this reaction with rabbit serum, then your goat anti-rabbit secondary antibody would recognize every rabbit serum protein in your sample, as well as the primary antibody you used. This would cause your sample to light up like a beacon and you would have many false positives. To minimize cross-reactivity issues, you can purchase serum that has been tested for cross-reactivity between multiple species or you can make sure that all your amplifying secondary antibodies are from the same host animal (e.g., you can use rabbit, mouse, and goat primary antibodies and use donkey anti-rabbit, donkey anti-mouse, and donkey anti-goat secondary antibodies to separate your results and use only one blocker).

Additionally, if your primary antibody is made in goat, then you shouldn't use nonfat dry milk, casein, or bovine serum albumin (BSA) in your blocking solution, because anti-goat antibodies also bind bovine proteins, and you get a lot of extra background noise in your blots or tissue samples. Similarly, if your primary antibody was generated in cow (they exist), then blocking with goat serum will generate high background noise. If you have a cow antibody and use cow proteins (milk, BSA, casein) to block your blot, then the whole blot will glow back at you after using an anti-cow secondary antibody. Additionally, milk is supposed to contain phosphoproteins, which can increase the background if you are using an anti-phosphoprotein primary antibody; this hasn't been my experience, but that likely depends on how good the antibody is.

The important part (for any antibody-based application) is that you perform your own antibody titrations. Few companies, in my experience, report optimal titers on datasheets. The issue at hand is not that I think companies are lying to me; on the contrary, I think they're telling the truth. Here's the big deal: the conditions that companies test their antibodies under are (very very likely) *not* the conditions you are testing their antibodies under. Many companies block their antibodies with BSA, for example; while that may work fine with cell lines and overexpression systems, BSA nearly always gives me an excess of noise and renders the experiment a waste of time. Some people like BSA, and if you're one of them, good luck to you, because I won't use it except as a last resort. In any case, you need to test your antibodies in your application to determine the right dilution (aka titer) to use. Titrations are discussed for each application in its own chapter.

Finally, antibodies are finicky biological entities. It can be difficult to determine the optimal parameters for an antibody in your particular application. You should start with the datasheet of the antibody from the vendor, as well as a copy (if available) of the vendor's protocol that was used to test the antibody. Here are the parameters that you can change to try to identify optimal conditions for the antibody:

1. *Detergent versus no detergent*: some antibodies work better without detergent. I typically test with detergent first.
2. *4°C, room temperature, or another temperature*: some antibodies work at room temperature, not 4°C, but can't be saved for later use, as they break down. However, since most antibodies work best at the body temperature of the animal, you get the most activity at 37°C and the highest background (in general). 4°C is the most

stringent treatment and generally gives lower background noise than leaving the antibody at higher temperatures.

3. *Optimal time to leave antibody with your tissue/membrane/cell line/tube*: this is especially important if you are testing antibodies at temperatures higher than 4°C. Less time means less binding but possibly less background noise as well. Additionally, this varies with your application, as noted in their respective chapters.

4. *Which detergent to use*: some antibodies work better in gentler detergents, and no one detergent is optimal for all techniques. If Tween-20 doesn't work, will Triton X-100? Will NP-40?

5. *Which blocker to use*: for some antibodies, BSA will block all the nonspecific binding sites. For others, goat serum will, but horse serum will not. Be willing to test multiple blockers.

6. *How much blocker to use*: if you are getting no signal, perhaps decreasing the concentration of blocker will help. If you are getting a lot of nonspecific background noise, increasing the concentration of blocker may fix the issue.

7. *Amount of antibody to use*: this is dependent on what application you are using the antibody for. Perform antibody titrations for each application, and possibly each blocker. Don't be afraid to dilute far past the manufacturer recommendations (more concentrated or less).

8. *How much detergent to use*: some antibodies work better with more detergent than others. Increase the detergent concentration to make the washes more stringent, decrease to make them less stringent.

9. *Which buffer to use*: again, this depends on your application. However, if there is a pH difference between your buffers, you may notice an effect on the antibodies.

There are times where going through this list would be prohibitive, in terms of monetary and time costs. As such, decide on how you are going to proceed at the start of your experiment. If the antibody doesn't work in your typical conditions, how many parameters will you test? Will you purchase new antibodies and hope to find one that fits your conditions? Some experiments will push you into this ("Two companies in the whole world sell antibodies against my target!"), and others will not. Decide at the start what the value of your time is versus the value of getting the antibody to work. If an immunoblot can make or break your experiment, and you paid thousands of dollars to make the antibody from scratch, it's worth your time to test all of these conditions. If it's not worth your time to test the conditions, then why did you spend the money? On the other hand, if you have three antibodies against your target in your refrigerator, and the first one doesn't work, test the other two before spending months on testing.

Antibody method selection

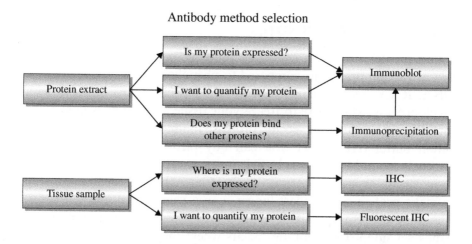

CHAPTER 6

Protein Extraction

PROTEIN EXTRACTION NOTES

I have greater success in collecting protein samples with specific protein extraction protocols, compared to phenol:chloroform extraction of DNA, RNA, and proteins. As with any dissection, this needs to be a quick process. Make sure to freeze your tissue as quickly as possible. I mix different aliquots of specific protease inhibitors into my homogenization buffer (HB); while you can buy plenty of protease inhibitor cocktails from different vendors, I find it cheaper and just as effective to make my own aliquots and use them later. I have not found a way to make protein samples last longer than a year in the −80°C freezer. The samples seem to start degrading after a year and immunoblotting bands get sloppier (generating smears where they did not exist before). While this is not the case with immunohistochemistry slices (more on that later), I don't know a comparable preservation method for unfixed, frozen tissue. As such, I plan my experiments and hope to collect protein samples for that experiment within a certain timeframe, so that I can do all the comparisons I hope to within a year.

I have used the centrifugation-based total protein and subcellular fractionation for mammalian cells and tissues. I include a detergent-based fractionation protocol in case someone needs it.

TOTAL CYTOSOLIC/MEMBRANE PROTEIN EXTRACTION PROTOCOL

1. Dissect tissue from animal and freeze immediately; store at −80°C until ready to extract. If collecting cells, trypsinize or scrape the wells and centrifuge cell/medium solution until cells form a pellet. Aspirate medium and store cell pellets at −80°C until ready to extract.

   I typically mix dry ice with 95% ethanol to make slurry. This is easier to handle than liquid nitrogen and typically cheaper. I have not had any problems with keeping my samples intact using this method. The dissection needs to be quick to avoid loss of samples.

Basic Molecular Protocols in Neuroscience: Tips, Tricks, and Pitfalls. DOI: http://dx.doi.org/10.1016/B978-0-12-801461-5.00006-X

2. Before extraction, prepare HB and mix with protease inhibitors (can also add phosphatase inhibitors and other inhibitors, depending on your protein targets). Keep on ice.

> I keep protease inhibitor stocks in the −20°C freezer and they keep for quite some time. Don't let the buffer/inhibitor solution sit too long; some reports indicate that these inhibitors lose activity after 30 min. There are multiple kinds of lysis buffer that different people recommend, depending on the kinds of proteins you're extracting; I haven't needed to use them, and you likely won't need to use them until you have a difficult target. As such, know that they exist (radioimmunoprecipitation assay (RIPA) buffer, NP-40 buffers), but don't worry about them until you have to.

3. Remove tissue/cell pellet from freezer and homogenize tissue, in HB with protease inhibitors, on ice.

> I typically use a Dounce 20 glass homogenizer with a fitted pestle for homogenization, because you can keep the glass cold; machine homogenization generates heat and destroys your proteins. I use just 500 μl of buffer to cover and homogenize the tissue, washing the pestle with another 500 μl of buffer afterward.

4. Transfer homogenate to a cold microcentrifuge tube.
5. Centrifuge samples at 6300g (more if you prefer) for 10 min at 4°C.

> Most protocols you read on total protein extraction refer to rpm instead of rcf (g), and so it's impossible to get their results (without knowing the model of centrifuge they're using). They also have high spin speeds that likely omit some cytosolic or membrane proteins from the final product, which is why I use this lower speed. Use a speed close to 6300g if you can't get that exact speed (I used 6500g on one centrifuge). This speed makes "cell debris" and the nucleus into a pellet; if you want to retain the nuclear fraction, lower the speed to 4000g.

6. Transfer supernatant to new tube, discard "cell debris" pellet. Measure concentration using the Bradford method and store at −80°C. Avoid freeze-thaw cycles.

> I don't use the spectrophotometer alone to examine the concentration of these samples because it vastly overestimates the amount of protein I have; this is possibly due to having so much extra "cell debris" that still absorbs the 280 nm light. I find that, after a year in the freezer, the samples still degrade; protein samples just don't keep as well as RNA or DNA. Storing samples in denaturing loading buffer seems to lead to faster degradation, in my experience. I list the Bradford method as a general suggestion; bicinchoninic acid (BCA) analysis works as another option.

"Cell debris" is one of those phrases you will find in most protocols, which is not defined. Your boss likely doesn't know what it is, and neither do I. I suspect the "debris" are simply parts of the nucleus that no one cares about.

SUBCELLULAR FRACTIONATION PROTOCOL (ADAPTED FROM TWO PAPERS)[8,9]

This protocol yields a pure cytosolic fraction, a nuclear fraction, and a membrane/organelle fraction. While the vesicles should be in the cytosolic fraction, the jury is still out on exactly how much or how little force is required to pellet the vesicles. To be safe, test a vesicle-specific protein on the fractions to determine what landed where if vesicle location/content is important for your experiment.

1. Dissect tissue from animal and freeze immediately; store at $-80°C$ until ready to extract. If collecting cells, trypsinize or scrape the wells and centrifuge cell/medium solution until cells form a pellet. Aspirate medium and store cell pellets at $-80°C$ until ready to extract.

    For the initial steps, my comments are the same as the above protocol; it's only at the spin steps that things change between these two protocols.

2. Before extraction, prepare HB and mix with protease inhibitors (can also add phosphatase inhibitors and other inhibitors, depending on your protein targets).
3. Remove tissue/cell pellet from freezer and homogenize tissue, in HB with protease inhibitors, on ice.
4. Transfer homogenate to a cold microcentrifuge tube.
5. Centrifuge samples at $6300g$ for 10 min at $4°C$.
6. Transfer supernatant to an ultracentrifuge tube, kept on ice.
7. Resuspend pellet with another half-volume of HB (that includes protease inhibitors) and thoroughly mix. Centrifuge these samples at $6300g$ for 10 min at $4°C$.

    This is likely the step that vastly increases my yield of membrane/organelle proteins. When I say "half volume," I mean that my total homogenate volume is 1 ml of solution, so I resuspend with another 500 μl of solution.

8. Transfer supernatant to the same ultracentrifuge tube.

9. Pellet remaining is the nuclear protein fraction that also has some cellular "debris." Keep if you want to examine nuclear proteins and store at −80°C.

> *If you want to examine nuclear proteins, you can resuspend the pellet and spin again at 4000g to pellet the nuclear fraction.*

10. Ultracentrifuge tube is centrifuged at 107,000g for 30 min at 4°C.
11. Transfer supernatant to a new tube. This is the cytosolic fraction; store at −80°C.

> *This fraction will be much more concentrated than any other fractions. Be prepared to dilute an aliquot of it in order to accurately assess the concentration, as reactions like the Bradford can be overwhelmed by high amounts of protein.*

12. Add a small amount of HB (depending on the size of the pellet) to the pellet in the ultracentrifugation tube. Keep on ice.

> *I have found no exact science to determine how much HB I put on the pellet. Start at a volume you're comfortable with and test it out. Keep volumes consistent.*

13. In a cold room (4°C), with samples on ice, use a sonicator at low amplitude (30% or so) to pulse (1 s on, 1 s off) the pellet/HB 10 times, then move to the next sample (wash sonicator with 70% ethanol between each use). Keep everything cold.
14. Repeat step 13 twice, for a total of three rounds of sonication. This should result in a resuspended pellet.
15. This is the membrane/organelle fraction. Store at −80°C.

> *While some people store their samples in loading buffer (different buffer than used in agarose gel electrophoresis), I find that this degrades my samples faster than if I save them in the HB that I've been using. I save my samples as is and mix them with loading buffer only just before I load samples into the acrylamide gel (see the Immunoblotting section in Chapter 7). Avoid freeze-thaw cycles and any frost-free freezers.*

16. Any of your fractions' concentrations can be measured using the Bradford or BCA methods.

> *As before, spectrophotometer readings of these fractions vastly overestimate how much protein is in the sample. I think that detergent-separated fractions work well in the spectrophotometer, but this protocol is not as clean.*

SUBCELLULAR FRACTIONATION PROTOCOL (DETERGENTS)

1. Dissect brain, rinse with **PBS**, and homogenize using a glass homogenizer (with fitted pestle). Homogenize on ice, using HB (HB2).
2. Divide homogenate between two microcentrifuge tubes. Add 1 ml of HB2.
3. Centrifuge for 15 min at 800g, 4°C.
4. Transfer supernatant to a new tube. Add 500 µl HB2 to each tube, resuspending the pellet. Centrifuge this resuspended pellet for 5 min at 800g, 4°C. Transfer supernatant to a new tube and save pellet (P1).

Fractionation, Enriched in Neurons and Glia
5. Combine supernatants, centrifuge for 5 min at 800g, 4°C. Transfer supernatant to new tube (S1).
6. Ultracentrifuge S1 for 45 min at 200,000g, 6°C.
7. Supernatant is cytosolic fraction; transfer to new tube.
8. Wash pellet with 1 ml of HB2, resuspend, and ultracentrifuge both pellet and cytosolic fraction at 200,000g for 45 min, at 6°C.
9. Transfer cytosolic fraction supernatant to a new tube and store this purified cytosolic fraction at −80°C.
10. Remove supernatant from pellet and add 1 ml of resuspension buffer plus protease inhibitors. Let sit 20 min.
11. Homogenize new pellet using glass homogenizer.
12. Add 110 µl of 10% Triton X-100 and mix well. Place this solution at 4°C for at least 2 h.
13. Remove an aliquot of this fraction (100−150 µl) and spin remainder for 45 min at 200,000g, 6°C.
14. Transfer the supernatant to a new tube (this is the detergent-soluble membrane fraction), and save the sample. Resuspend pellet in 1 ml of resuspension buffer, homogenize, and add Triton X-100 to 1%.
15. Spin detergent-soluble and detergent-insoluble membrane fractions for 30 min at 200,000g, 6°C.
16. Remove detergent-soluble membrane supernatant and save at −80°C. Resuspend detergent-insoluble pellet in 1 ml resuspension buffer, add Triton X-100 to 1%, and store at −80°C.

Fractionation of P1 Pellet (Microvessel-Enriched Fraction)
17. Resuspend P1 pellet in 2 ml of resuspension buffer + dextran and mix well with pipet.

18. Centrifuge for 15 min at 4400g, 4°C.
19. Remove supernatant. Resuspend and wash pellet with 500 μl of resuspension buffer + dextran.
20. Centrifuge for 5 min at 4400g, 4°C.
21. Resuspend P1 pellet in 1 ml of resuspension buffer with protease inhibitors and homogenize using glass homogenizer.
22. Centrifuge for 5 min at 4400g, 4°C.
23. Repeat step 21 and sonicate on ice for 10 pulses (1 s on, 1 s off). Add Triton X-100 to 1% total. Leave at 4°C overnight.
24. Remove 100 μl of this fraction and spin the remainder for 45 min at 200,000g, 6°C.
25. Transfer the supernatant to a new tube (this is the detergent-soluble fraction) and save the sample. Resuspend pellet in 1 ml of resuspension buffer, homogenize, and add Triton X-100 to 1%. Spin supernatant and pellet for 30 min at 200,000g, 6°C.
26. Save supernatant of detergent-soluble fraction at −80°C. Resuspend pellet in 1 ml of resuspension buffer, homogenize, add Triton X-100 to 1%, and store at −80°C.

SOLUTION RECIPES

Homogenization buffer (HB)

Sucrose	54.77 g
Tris base	0.61 g
KCl	1.86 g
EDTA	0.15 g
dH$_2$O	To 500 ml

pH to 7.8, vacuum filter.

Aprotinin stock

Aprotinin	2 mg
dH$_2$O	1 ml

Aliquot, store at –20°C, do not reuse after defrost.
Inhibits trypsin, chymotrypsin, and plasmin proteases.

Leupeptin stock

Leupeptin	10 mg
dH$_2$O	1 ml

Aliquot, store at –20°C, do not reuse after defrost.
Inhibits lysosomal proteases.

Sodium fluoride stock

NaF	0.42 mg
dH$_2$O	10 ml

Aliquot, store at –20°C, don't reuse after defrost.
Inhibits serine and threonine phosphatases.

Resuspension buffer

HEPES, pH 7.4	25 mM
EDTA, pH 8.0	2 mM

Add protease inhibitors, as noted above (HB).
Can add dextran to 17.5%.

Homogenization buffer + inhibitors

HB	varies
Aprotinin	2 µg/ml
Leupeptin	10 µg/ml
Pepstatin A	1 µg/ml
PMSF	1 mM
NaF	10 mM
NaVO$_4$	1 mM

*Use only the inhibitors you need for the experiment.
Don't use NaF or NaVO$_4$ if you aren't examining
phosphoproteins.*

Pepstatin A stock

1 ml	1 mg
Methanol	1 ml

Aliquot, store at –20°C. Can reuse stock.
Inhibits aspartic proteases.

Phenylmethanesulfonylfluoride (PMSF)

PMSF	1.8 g
100% ethanol	10 ml

Aliquot, store at –20°C.
*PMSF does not dissolve in ethanol, so cut your pipette
tip to allow the PMSF into the tip. Inhibits serine and
threonine proteases.*

Homogenization buffer 2 (HB2)

HEPES, pH 7.4	10 mM
Sucrose	300 mM
EDTA, pH 8.0	2 mM

Add protease inhibitors, as noted above (HB).

Sodium orthovanadate stock

NaVO$_4$	1.84 g
dH$_2$O	10 ml

Inhibits tyrosine phosphatases.
Set pH to 9.0 using HCl, boil until colorless. Cool to room temperature, pH to 9.0 again.
Boil again until colorless, repeat pH adjustment, and boiling until pH remains at 9.0 after boiling and cooling.
Aliquot, store at –20°C, do not reuse after defrost. Discard if aliquots turn yellow.
And now you see why most labs get phosphatase inhibitor capsules that dissolve in HB.

CHAPTER 7

Immunoblotting (Western Blot)

IMMUNOBLOTTING NOTES

Immunoblotting is the process of using antibodies to identify proteins that have been transferred onto a membrane. Immunoblotting is also called western blotting and was named by analogy; Dr. Edwin Southern invented Southern blotting (DNA on a membrane), but northern blotting (RNA on a membrane) was named to go along with the name of the first, and western blotting (protein on a membrane) was named to go along with the other two. As such, I was admonished by my English teacher that Southern blotting is capitalized because it's the guy's name, while the others are not capitalized (because they're just attempts by scientists to show that we're witty). I have performed both chemiluminescent and fluorescent immunoblotting, which have different caveats and treatments. There are many great resources for immunoblotting, including the Abcam Protocols Book,[10] the Thermo Scientific Western Blotting Handbook,[11] and many online troubleshooting guides.

Immunoblotting, in addition to identifying the presence of a protein, is also able to identify changes in protein concentration between samples, the quality of which depends on the system you use to examine those changes. Immunoblotting is known as a semiquantitative analysis, because there is a great deal of variability between blots (transfer efficiency, loading the exact same amount of protein per well, wash times, and the exposure of the film) and because you are not comparing your proteins to a standard curve of the protein itself (which would help with absolute quantitation). We typically perform relative quantitation of blots in a manner similar to quantitative PCR (qPCR): how much did the target change relative to my reference proteins (aka loading controls)?

The argument can be made that fluorescent immunoblot is more quantitative and sensitive than chemiluminescence (in practice). The reason that chemiluminescence is usually regarded as more sensitive than fluorescence is that it works by enzymatic action; if the enzyme is present, you'll eventually see some reaction product. If altering the

Basic Molecular Protocols in Neuroscience: Tips, Tricks, and Pitfalls. DOI: http://dx.doi.org/10.1016/B978-0-12-801461-5.00007-1

gain doesn't let you see the fluorescence, even if there's a small amount present, you won't see that fluorescence. However, the issue with chemiluminescence is that you are not only dealing with the enzyme and any enzymatic issues like running out of substrate, you are also dealing with your data acquisition method. The most common method, autoradiography film, has a minimum amount of light required before it will show a band and has a maximum amount of light that it can handle before all the bands look the same. Additionally, light can scatter in the film and lead to large blotches on your film. Most imagers (those developed for chemiluminescence or fluorescence) have a greater dynamic range than film, meaning that they should be able to pick up and resolve signals of greater and smaller amounts of light. Near-infrared fluorescent immunoblotting (using the LI-COR® Odyssey®[1] Imaging System) avoids background light from the visual range of light and so, by decreasing background, allows for the use of less protein in your blots. For some antibodies, fluorescent blotting works fine, but for others, chemiluminescence was the only way to show immunoreactivity, so test each antibody individually. Colorimetric blotting is also possible (forming a colored precipitate at your band), but is not quantitative at all, and so you won't see it much, due to the difficulty of objectively determining whether your band is more blue/brown/green/etc. than the next band.

Most chemiluminescent reactions involve secondary antibodies conjugated to horseradish peroxidase (HRP), which reacts with hydrogen peroxide and luminol, releasing light. Some kits use reagents that react with alkaline phosphatase (AP), but those are less common than HRP. There doesn't appear to be a big difference between HRP and AP results, so use what is available to you.

Analysis of your data will depend on what method of blotting you perform. Fluorescent imagers for blotting typically have their own densitometry analysis software that comes with the machine. Chemiluminescence reactions involve film or cooled charge-coupled device (CCD) cameras. Film requires you to be fast, as the enzyme is working as soon as it has substrate to use, and when the substrate is gone (luminol, in this case), the data go with it. Whenever you have a picture of some sort, you can use densitometry software available from different vendors (including NIH's ImageJ). If you use film, you will

---

[1]LI-COR® and Odyssey® are registered trademarks of LI-COR Biosciences.

need to scan your film to generate a computer image. Office scanners, while commonly used, are not good for your data.[12] Turn off any automatic gain controls for your scanner of choice.

Stripping and reprobing with a second antibody is standard fare for immunoblotting, as most experimenters want to examine their protein of interest and a "loading control." The loading control is a bit of a misnomer; it is a check for sample degradation, but it doesn't necessarily adjust for any differences in your pipetting (which, if left unadjusted, can alter your quantitative data) and, like a reference gene for qPCR, is something that does not change due to your experiment. To accurately assess if you loaded the same amount of protein, you can simply stain your blot with Ponceau S or Fast Green, take a picture, and analyze the image using densitometry (see protocols[13] on page 69). With chemiluminescence, you can probe for both your protein of interest and your degradation control at once if they are at different weights; you can cut your blot into two pieces, separated by weight, and probe them separately. You can't put them together because both will require HRP to give a result, and the bands will be indistinguishable. This is the main advantage of fluorescent immunoblotting (other than quantitation): if I make my protein(s) of interest different colors (via secondary antibody conjugates), then I can examine them all at the same time. The LI-COR Odyssey® uses two colors in the infrared spectrum, which cuts down on background fluorescence at the expense of three-color (or more) multiplexing; laser scanners like the GE Typhoon™[2] can examine three colors at once (or more, for other models), but those are in the visible spectrum, and generally have higher background fluorescence. When planning any fluorescence experiment, make sure that your chosen fluorophores do not have overlapping emission spectra; if they do, you won't be able to separate target A from target B due to bleed-over artifacts.

Nitrocellulose membranes interestingly give low background for chemiluminescence and infrared fluorescent blotting, but high background for visible spectrum fluorescence. Polyvinylidene fluoride (PVDF) membranes are more expensive and have superior protein binding when compared to nitrocellulose membranes, but have high background in fluorescent blotting unless you purchase the low-background PVDF (like Millipore's Immobilon-FL™).

[2]Typhoon is a trademark of Amersham Biosciences Limited.

One important note: I have tested some antibodies that simply do not show up on fluorescent blots that do show up on chemiluminescent blots. This is likely due to differences in buffers and/or sensitivity; if you don't get results in fluorescent blotting, give chemiluminescent blotting a try (in addition, antibodies are just funny; one anti-β-tubulin antibody that worked wonders for fluorescence took 30 min to show up via chemiluminescence).

The process of immunoblotting begins with protein extraction (detailed previously), followed by the casting of a polyacrylamide gel and subsequent electrophoresis (separating proteins in your samples by size; also called SDS-PAGE), the transfer of proteins from the gel to a membrane, and subsequent blocking and treatment steps. But before running the experiment, you need to determine how much antibody to use. In immunoblotting, the antibody titer you use depends on a couple of factors:

1. Are you using chemiluminescence or fluorescence to visualize your data? In practice, I use different titers for each of these techniques.
2. What amount of protein are you using?
3. Where did the protein sample come from? Is your target expressed in high or low abundance in that particular tissue?

For immunoblotting, I have a less refined method than for immunohistochemistry to titrate antibodies. I use the manufacturer's recommendations as a middle point, test titers both more and less dilute than that recommendation, and compare the pictures. After that test, I use the most dilute amount I can that still gets a good picture (this can also be done using a technique called a Dot Blot; see protocol on page 70). Typically, I combine the antibody titration with the protein concentration titration. I set up a blot like this:

Ladder	5 µg	15 µg	30 µg	5 µg	15 µg	30 µg

This gives two repeats to account for any differences in my pipetting. If you like, you can simply split this between different antibody concentrations. This helps identify the minimum amount of protein and antibody for your experiment. I use the minimal amount of protein and antibody for reasons of both cost and quantitation; if the X-ray film gets overexposed or fluorescence reaches the limit of detection, you can't identify any changes, much less small ones that are still biologically relevant. You should perform secondary antibody titrations as well; secondary antibodies can increase background if you have too high a titer. For a secondary antibody titration, run a blot with different amounts of protein but no primary antibody. Test dilutions of the secondary until you completely lack background noise. You can then test again with a primary antibody to determine the lowest titer you can use that gives satisfactory results.

While you can use a biotinylated secondary and HRP- or fluorophore-conjugated streptavidin for additional amplification of the signal, this is uncommon in immunoblotting practice (but can still be done, after washing off the secondary antibody). Use of secondary antibodies without other amplification is much more common in immunoblotting.

So after all that explanation about planning your experiment, what are you actually doing? You start with a different kind of gel electrophoresis, sodium dodecyl sulfate polyacrylamide gel electrophoresis (SDS-PAGE). Proteins don't work so well in agarose gels, so we use polyacrylamide gels instead. Proteins also have varied charges, so we apply SDS, which binds to the proteins and gives them a uniform negative charge. Similarly to agarose gel electrophoresis, we can separate the proteins by size. SDS, combined with either β-mercaptoethanol or DTT, partially denature our proteins so that they become linear (heat is necessary for full denaturation with β-mercaptoethanol but not necessarily so for DTT). We separate our denatured proteins by size and transfer the proteins from the gel to a membrane that binds proteins (nitrocellulose or PVDF). We can then treat that membrane with antibodies to detect our protein of interest (the newer, in-gel immunoblotting methods are not discussed in this book).

SDS-PAGE AND IMMUNOBLOT PROTOCOLS

Use ultrapure water for each solution, as normal distilled water gives high background on blots. Handle blots with forceps, as fingerprints show up on blots.

SDS-PAGE and Membrane Transfer Protocol

1. Plan the entirety of your experiment first; you can run some steps or prepare multiple solutions simultaneously to save time.

   While this is true for every experiment, this is the difference between spending 6 or 10 h in a day on getting an immunoblot experiment started. This is also a good time to check if the X-ray film developer is low on any developer or fixer solutions, if you need that machine.

2. Prepare acrylamide gel solutions (both running gel and stacking gel) without APS (see Solutions) or TEMED.

   In my experience, I can prepare my gel solutions ahead of time and they stay good for at least a month when I omit the APS and TEMED; they would probably last longer than a month. The key is that APS in the gel without TEMED will cause your gel to polymerize overnight, and TEMED in the gel without APS would likely lose activity over time. So, if you have a day where you're unable to run your immunoblots but have free time, you can get the solutions ready in order to cast gels immediately. Store them in sealed containers at room temperature. If you need to use an antibody that does not recognize denatured epitopes, you can omit SDS and make it a "nondenaturing PAGE."

3. Prepare gel casters; this will change depending on the equipment being used.
4. Add APS and TEMED to running gel solution, mix, and pour gel into caster. Add a small amount of isopropanol to the top of the gel to keep the top of the gel a consistent height and to remove bubbles.

   I prefer to use a micropipettor to move gel solution between the plates of the caster. A Pasteur pipette also works well.

5. Wait for gel to set. The time this takes is dependent on the size of the gel and the amount of APS added.

   Increasing the amount of APS in solution results in faster gel polymerization. If you are using a minigel system, including 25% APS helps polymerize gels fast and gets you started faster, but it is too fast for older, larger gel systems (which could polymerize unevenly and your blots would look odd). For older systems, use of 10% APS allows even polymerization of gels.

6. Pour out/remove isopropanol layer.
7. Add APS and TEMED to stacking gel, mix, and pour gel into caster. Add well caster.

Evaluate the amount of stacking gel you make as you practice the method. Many suggested volumes for two gels can easily be used for up to four gels.

8. Wait for gel to set. While gel is setting, prepare samples and mix with loading buffer.

9. Depending on the experiment, heat the samples in a 95°C bath for 5 min and immediately put on ice afterward. Leave on ice until the gel is ready for sample loading. Do not let samples boil.

This is an important step. β-mercaptoethanol and SDS work in concert to denature your proteins in this solution (so that they will separate by size in SDS-PAGE, not by 3-D structure or charge), but they don't completely denature proteins by themselves—they need heat or another chemical denaturant to work completely (DTT, an alternative to β-mercaptoethanol, may work without heat but loses activity over time while in solution).[14] Whether you heat your samples or not depends on the antigen you're looking for; some proteins precipitate out of solution in the presence of heat, and so you might use loading buffer with urea to fully denature without heat (semidry transfer generates some heat, so that can't be used if you want to avoid heating the samples). Some antibodies have target antigens that cannot be denatured without a loss of signal; use typical loading buffer without heat to see those. If you have urea in the sample, don't heat it. The urea will break down into a couple of caustic materials and that means your samples won't survive the process. Finally, don't boil the sample. Proteins aggregate in boiled samples and won't be fully separated by the SDS-PAGE, altering band weights.

10. Remove gel from casting apparatus and put into electrophoresis apparatus and tank. Cover in running buffer to whatever point your apparatus requires.

Make sure to cover the inner wire with buffer and that the level of liquid is above your wells (unless otherwise stated by the manufacturer); if you don't, the circuit will be incomplete and your samples will not move in the gel.

11. Load samples and ladder into wells.

The choice of ladder isn't a huge issue, but I prefer ladders where each size band has an individual color. The reason for this is that, if some bands don't transfer well, you can still identify what sizes of proteins did transfer. Single- or two-color ladders can be confusing as to what actually transferred ("Is this the first or second time that I should see blue?").

12. Select amperage, voltage, and time to run gel.

> *Like agarose gel electrophoresis, changing voltage and amperage alters the resolution of your gel and how pretty it looks. I typically run at 100 V for 90 min and this gives me fine results. I have used high-voltage buffers (commercially available) that can be run above 150 V (time shortens from 3 h to 45 min for gel running and transfer). They work well, but only for samples that can stand heat. The solutions are also expensive if you are running many blots.*

13. When the gel is done (blue stain is at the bottom of the gel or has run off), transfer gel to a tray of blotting buffer (also called transfer buffer). If using a minigel system, move directly to the next step. If using an older system with larger gels, leave gel in blotting buffer for 30 min to allow the gel to shrink.

> *If using PVDF membranes, treat the membrane with methanol at this time to get it ready (it will turn an opaque grey color and stands out very much against the white, inactivated PVDF). Follow manufacturer instructions on "activating" PVDF. By leaving the gel in blotting buffer, you allow the gel to shrink in order to use less membrane per experiment.*

14. Set up transfer apparatus according to manufacturer instructions. I will describe the process of a "full-wet" transfer; semidry transfer systems, among others, follow slightly different instructions. Make a "sandwich" in the transfer apparatus of a sponge (comes with the apparatus or can be ordered from the manufacturer), filter paper, gel, membrane (nitrocellulose or PVDF; see Notes), filter paper, and sponge. Make sure that the membrane is on the cathode end.

> *Full-wet transfer is useful to know. While different companies sell devices that allow for faster transfers, they may not work for all experiments (especially if they generate excess heat). The classic transfer method for high-weight proteins is to run the transfer on a low voltage overnight and come back in the morning. I'm not patient enough to make a 3 + day experiment into a 4 + day experiment, and full-wet transfer usually works great for proteins large and small. If you are looking for larger proteins (greater than 110 kDa), you can increase the amount of SDS and decrease the amount of methanol in the blotting buffer.*

15. Set up "sandwich" in holder and put into the tank. Fill tank with blotting buffer (just above the "sandwich" in order to keep

the circuit complete) and cover. Attach electrodes and, if transferring in a noncold room (not 4°C), put entire apparatus into an autoclave tray (something that won't leak) and cover with ice: each side and on top (really, cover the whole apparatus with ice).

Some people, alternatively, run their transfer in a 4°C room, leaving the transfer apparatus on a magnetic plate with a stir bar at the bottom of the transfer tank. This allows the cold fluid to circulate and disperses the heat; I find that covering with ice works better for my blots. Test each to determine which one you like best. The important thing is that you have some way to mitigate the heat.

16. I set my transfers for 90 min at 100 V. The timing and voltage can be changed depending on the size of your target, but this has worked for proteins from 110 to 10 kDa (I haven't examined anything smaller).

17. When the transfer is done, remove it from the apparatus and mark it with a colored pencil (gently!). At this point, you can stain the gel and/or the membrane to check if your transfer worked. From this point and onward, always handle the gel with tweezers (fingerprints show up as fingerprint-shaped noise on blots).

I mark the blot because stripping the blot usually washes off some of the dyes in my marker lane but doesn't wash away colored pencil markings as effectively. This way, I always know which end is supposed to be up. To check the success of your transfer, gels can be stained with Coomassie Blue, while membranes can be stained with Fast Green (nitrocellulose) or Ponceau S (PVDF). If you are determining the optimal time and settings for your transfers, these staining steps are crucial.

Immunoblotting Protocol

18. Wash the membrane briefly (less than a minute) in ultrapure water.

19. Put the blot into blocking buffer. Blocking buffer is one of the more important aspects of your experiments, so plan this part carefully. Block for 45–60 min.

Blots can be blocked with nonfat dry milk, BSA, casein, detergent, normal serum (same species as the secondary antibody), or commercial blockers. I find that milk is cheap and gives low background in chemiluminescence but not near-infrared fluorescence. BSA gives me high background no matter what I do (except one Cell Signaling Technology antibody, using infrared fluorescence). I don't have experience with casein blocking or animal-free blocking (just buffer and detergent). Background noise with serum block varies for chemiluminescence but I don't have problems with fluorescence. I don't like to use commercial blockers because I don't know what is in them.

I mix the blocker with PBS (fluorescence) or TBS (chemiluminescence) with detergent, and the antibodies are always mixed in the same solution (e.g., block with 4% milk/TBS + Tween (TBST), primary antibody is in 4% milk/ TBST, secondary antibody is in 4% milk/TBST). Test different blockers to determine which is optimal for your antibody and your protocol (fluorescence versus chemiluminescence). The amount of solution you make depends on the size of your blot and the container you keep the blot in; I typically choose boxes where the blot is completely covered by 8 ml of solution.

20. When the block is done, dilute the antibody in solution (same solution as the blocking buffer, except including detergent), remove the block buffer, and replace with the antibody. I leave this on a shaker overnight at 4°C; some antibodies work better at different temperatures. I perform a type of incubation called "floating a blot," which I got from the BiteSizeBio.com website (see figure on next page).

I don't save block solutions unless they come from a commercial supplier, and I've tested them to ensure that I don't get excess background by doing so.

To float a blot, you need Parafilm™ (waxy film) and airtight Tupperware. Put a little blocking buffer on the bottom of the Tupperware and spread it around. Then put Parafilm™, cut to a little larger than the blot, in the Tupperware (the liquid "seals" the Parafilm™ to the bottom). I dry up excess blocking buffer (you can also use water or TBST with no extra background). When ready, I mix the antibody in 1 ml of solution instead of the 8 or so I would need if I did not float the blot. Pipet the 1 ml of antibody solution onto the Parafilm™ and place the blot face-down onto the antibody solution (adjust volume if you have smaller or larger membranes). Make sure that the antibody solution has spread across the entire blot. Cover the Tupperware™ (needs to be airtight to avoid evaporation) and leave on a shaker at 4°C overnight. This saves a great deal of antibody if you can't save the solution for later (and I've never had a time with chemiluminescence where I could save the antibody; milk is mold food). I have never had an increase in background when using this method; I can't imagine switching back.

(1) Air-tight container with lid

(2) Wet bottom with buffer

(3) Make sure that a thin
layer of buffer is spread
throughout the container bottom

(4) Put waxy paper onto bottom;
will stick to bottom because of
the buffer

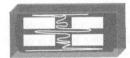

(5) Wipe up excess buffer;
put antibody/buffer onto
waxy paper

(6) Lay blot face-down onto
waxy paper, making contact
with antibody/buffer

(7) Seal with lid, treat as normal!

21. The next day, remove the antibody solution from the blot. Save antibody solution if desired.

> *Earlier, I mentioned that chemiluminescent reactions featuring HRP were incompatible with sodium azide. Fluorescent reactions are not, and so you can load all your antibody solutions with sodium azide with no loss of data. I have been able to reuse antibodies for fluorescence up to five times with no loss of data (adjust the gain); I haven't combined fluorescence with floating blots, but that's because it doesn't seem necessary when I'm saving the antibodies for five different blots.*

22. Three washes, 10 min each, in **PBST** (fluorescence) or **TBST** (chemiluminescence).

> *PBST gives lower background for near-infrared fluorescent blotting, while TBST gives low background for chemiluminescence. Avoid PBST if using any AP reagents.*

23. Remove solution and replace with secondary antibody (in blocking buffer solution), let sit at room temperature for 90 min (cover if performing fluorescent blotting).

While I've found that the overhead fluorescent lights don't affect my near-infrared fluorescent blots (and thus covering is unnecessary), that is likely dependent solely on the brand and type of light bulbs used; since you and I won't be told when maintenance decides to change brands, just keep the blots covered anyway.

24. Three washes, 10 min each, in **PBST** or **TBST** (if you covered it before, keep it covered).

    This is a good time to set up for the next step, including turning on any X-ray film scanners and letting them warm up to operating temperature.

Fluorescence

25. Keep protected from light for the remainder of the process. You can take the blot immediately to your image scanner to see your blot. You may also dry out the blot to increase the signal to noise ratio, but if you dry it out, you can't strip and reprobe the blot.

    Membranes with near-infrared fluorophores are supposed to last indefinitely if dried out. Visible spectrum fluorophores likely last a long time, as well.

26. Use the software from your imager to collect a picture and perform any densitometry required.
27. When finished, get blot ready for stripping and reprobing, if necessary.

Chemiluminescence

28. Set up X-ray film developer (or digital scanner, if you've got the money for the latest gadgetry), X-ray film, a marker to mark each film, scissors, an X-ray film case, timer, and chemiluminescence reaction solutions. Get reaction solutions ready absolutely last; the peroxide solution is degrading as soon as you expose it to light.

    Most X-ray film developers have specific temperatures that they need to reach before they work efficiently. Some people just put their film in while the machine is cold and just turned on. I don't want to be in a situation where I have no data from a day and a half of work, and I have to wonder if it was the experiment or if I wasn't patient enough with the machine. Just wait and use the machine properly—it won't kill you. Additionally, you can make your own chemiluminescence reaction solutions, but most people buy them. Picogram-sensitive solutions are good for quantitation, while femtogram-sensitive solutions are so strong that they're typically only good for a yes or no (I once had the whole blot glowing at me while using the stuff—every last inch of it).

29. The X-ray film case requires a little extra setup. A clear "blot holder" needs to be assembled, and typically I've used the clear, plastic paper holders, cut to fit blots, and held to the case by tape.

   You want to avoid getting luminol solution directly on your film, which is why we make this cover. The clear plastic paper holders can be purchased at any office supply store. Make sure they're larger than the blots you want them to hold.

30. Put the blot, face up, onto something that can get wet. Mix both portions of your chemiluminescent reaction solution multiple times, pour this mixture onto your blot, and leave the mixture on the blot for 1 min.

31. Pick up the blot with tweezers and let the excess fluid drain off. Put the blot into the "blot holder" you made, press all the air bubbles out (if they are present, they'll show up on your film and you won't see your data). Turn off the lights, leaving only darkroom lights (red) on.

   Do not open your box of X-ray film while the lights are on. You'll ruin the film if it gets exposed to the regular lights.

32. Cut the corner off a piece of X-ray film (so you know which way is "up" and which side is correct) and place against the outside of your "blot holder." Close the X-ray film case and lock it (if it has a lock mechanism; if not, press gently but firmly on the case to make sure that your film is close to the blot). Let it sit for some specified time; if I don't know the antibody very well, I start at 1 min and move onward from there. Alternatively, a CCD camera or specialized "blot scanner" can be used to collect pictures.

   You can judge where to go by the look of the 1 min exposure; need just a little more, do a 3 min exposure. Do you need a little less? Count for 45 s. Don't see anything? Try 5–10 min. If you want to make a quantitative comparison, the optimal exposure gives you full bands and still allows you to see a difference between bands (not overexposed).

33. When the time is up, remove the film from the case (gently; make sure that you don't take the blot with it), and place into the warmed-up and ready X-ray film developer.

34. When your film comes out, you have some initial data. Collect additional films for longer periods of time as necessary. Typically, the HRP-conjugated secondary antibody has an "enzyme life" of

4 h; if I don't get signal otherwise, I leave some film on the blot for the remaining time or overnight. This will generate lots of background, but can give me a signal if nothing else will. While not figure worthy, it gives me a place to go.

35. When finished, get blot ready for stripping and reprobing, if necessary. Allow film to dry. Data analysis can be done anytime, but you will need to scan the films (make them into computer image files) and analyze the densitometry of the images using a program like NIH's ImageJ. Specialized scanners perform this function best instead of standard office scanners.[12]

PVDF Blot Stripping Protocol[15]

Stripping a blot can be done in different ways, and these can depend on your choice of blotting membrane. For PVDF membranes, I chose a blotting buffer initially described by Yeung and Stanley,[15] and for nitrocellulose, I have used commercial stripping buffers as well as homemade buffers. If you get a commercial buffer, follow their directions. Citrate and acid stripping buffers, present in some older protocols, take 2−4 h to strip and are not noticeably better than the commercial buffers that take 15−30 min. Note that any stripping protocol will remove some of your protein as well as the antibody complex. I will describe the Yeung and Stanley protocol here (do not use for nitrocellulose, because the membrane will fall apart).

1. Wash buffer in TBST for 5 min.
2. Add β-mercaptoethanol to the stripping buffer.
3. Four washes of stripping buffer, 3 min each.

   The smell of β-mercaptoethanol is strong, so perform these washes under a fume hood, if possible.

4. Six washes of stripping wash buffer, 5 min each.
5. Optionally, redevelop membrane to examine if there is any immunoreactivity left. If you do this, wash another four times, 5 min each, in TBST to wash ECL solution/antibodies off membrane.
6. Incubate with the next antibody. Reblock if desired.

   I have not found reblocking to be necessary. You can restain with Fast Green or Ponceau S to determine how much protein was removed by your stripping procedure.

Blot Stripping Protocol (Alternative Stripping Buffer)[16]

This protocol can be used for both nitrocellulose and PVDF membranes. Remember, any stripping protocol will remove some of your protein as well as the antibody complex.

1. Four washes of TBST, 5 min each.
2. Leave membrane in Alternative Stripping Buffer for 30 min at 50°C with agitation.
3. Six washes of TBST, 5 min each.
4. Optionally, redevelop membrane to examine if there is any immunoreactivity left. If you do this, wash another four times, 5 min each, in TBST to wash ECL solution/antibodies off membrane.
5. Incubate with the next antibody. Reblock if desired.

   You can restain with Fast Green or Ponceau S to determine how much protein was stripped from the membrane.

COOMASSIE BLUE PROTOCOL

1. Apply 0.5% Coomassie Blue G-250 to cover the gel. Stain for 5 min.
2. Discard stain and wash with ultrapure water briefly.
3. Wash in Coomassie Blue destain solution for 10 min. Replace solution until discrete bands are visible (can take hours; I typically leave the gel in ultrapure water overnight at 4°C).

FAST GREEN/PONCEAU S PROTOCOL

1. Apply stain to membrane and let sit for 15 min (Fast Green) or 1 min (Ponceau S) on a shaker.
2. Pour stain back into stock bottle (can be reused) and apply destain solution (Fast Green) or ultrapure water (Ponceau S).
3. Wash in destain (Fast Green) or ultrapure water (Ponceau S) for 10 min at a time. Perform additional washes as necessary, until you see discrete bands.
4. Take a picture of the blot. You can analyze this picture using densitometry to determine if you loaded a consistent amount of protein across your lanes.

DOT BLOT PROTOCOL[17-19]

1. Prepare dilutions of protein samples in TBS or PBS. Test a wide range of dilutions.
2. Prepare membrane strips (nitrocellulose or PVDF) and label each strip with a pencil for its primary and secondary antibody concentrations.

   Typically, one or two dilutions of primary antibody are tested with two to three dilutions of secondary antibody. If using PVDF, activate the membrane with methanol.

3. Place membrane strips on a dry paper towel. Pipet protein samples onto strip (they will appear as dots) and allow to dry for 10–15 min, until no moisture is visible.
4. Block strips in blocking buffer of choice for 1 h, room temperature, with shaking.
5. Prepare dilutions of primary antibody and apply to membrane strips. Incubate for 1 h at room temperature or overnight at 4°C.
6. Wash membrane strips 4 × 5 min in TBST/PBST.
7. Add secondary antibody to membrane strips for 1 h, room temperature, with shaking.
8. 4 × 5 min in TBST/PBST.
9. Develop pictures via fluorescence or chemiluminescence.

IMMUNOBLOT CONTROLS

1. *No primary antibody control*: tests for nonspecific binding of the secondary antibody to blocking reagents, different proteins in the membrane, or the membrane itself (anti-chicken secondaries bind spuriously to PVDF membranes, for instance).
2. *Negative controls*: protein samples that should not express your protein of interest (knockout animal tissue, cell lines, tissues that don't express your protein normally) are useful to identify antibody specificity.
3. *Positive controls*: protein samples that do express your protein of interest are useful to identify antibody or blocking problems. Can also use a sample that is known to be affected by your experiment to ensure that your experimental treatment went as planned.
4. *Peptide preincubation control*: this control is accomplished by incubating the antibody with the antigenic peptide before

applying it to the immunoblot. This control is problematic. Briefly, if staining disappears after preincubation, it could mean that the antibody is specific or that the antibody has greater affinity for the free-floating peptide than for proteins on the membrane (and membrane staining may still be nonspecific). If staining remains after preincubation, the antibody may either be nonspecific or have greater affinity for the protein bound to the membrane than for the free-floating peptide. Combine with other controls.

5. *Single-color controls*: for multiplex immunoblotting (fluorescence), single-color/single-antibody controls can identify if your antibodies are binding to each other instead of their intended protein targets. This can also identify if you are getting bleed-over artifacts from the fluorophores you are using.

6. *Omit primary and secondary antibodies as a control*: only useful if you use a tertiary component to amplify the signal (such as streptavidin). Identifies spurious binding or noise coming from your tertiary component. Unnecessary if omitting the primary antibody alone eliminates the signal.

7. *Multiple antibodies against different parts of the target*: an expensive control, but useful. If you use different antibodies that target different parts of the same protein, you can compare bands. If the bands are the same, that's a good indication that your antibodies are specific and targeting the same protein. It's not infallible, so combine with other controls.

8. *Use antibody in immunohistochemistry to validate antibody*: this control can be useful but has some drawbacks. This control is considered a success if your immunoblot has bands at the correct weight(s) and your immunohistochemistry shows staining in the right places. However, immunoblotting and immunohistochemistry use antibodies to identify targets in very different scenarios: immunoblotting examines a protein fixed to a membrane, while immunohistochemistry examines a protein fixed within tissue. Many antibodies that work for one technique don't work for the other, which can give a false negative. Further, antibodies that are specific for one technique can have spurious binding when used in the other technique. This control can also take a lot of time, as you would need to perform controls for both techniques in order to truly assess specificity. While this can suggest specificity, it is best combined with other controls.

IMMUNOBLOT TROUBLESHOOTING[10,20]

Issue	Possible Cause	Possible Solution
High background:	Dirty transfer pads (sponges) or tanks.	*Wash out all tanks and equipment using dH₂O. Allow to dry completely and check for crystallization.*
	Acrylamide residue left on membrane.	*Remove acrylamide from blotting membrane as soon as transfer is complete and wash all tanks and equipment thoroughly.*
	Bacterial/mold growth in solutions.	*Make new solutions from scratch and test a new blot.*
	Fingerprints.	*Don't touch the membrane directly, even with gloves.*
	Dirty forceps.	*Wash forceps thoroughly with distilled water.*
	Insufficient blocking.	*Test both a longer blocking time and increase percentage of blocking reagents (from 1% to 5% milk, for example).*
	Secondary antibody issues.	*Dilute secondary antibody in your blocker of choice, testing dilutions until nonspecific background disappears. Change blocking reagents, if necessary.*
	ECL prepared poorly.	*Prepare ECL just before use.*
	Antibody cross-reactivity with blocking solution.	*This should be identified during antibody titrations. Change blocking reagents.*
	Insufficient washing.	*Increase wash time or number of washes.*
	Exposure to film was too long.	*Change film exposure times; decrease exposure time to decrease background.*
	Antibody cross-reactivity.	*Only an issue for fluorescence; test each antibody by itself to determine if antibodies are binding to each other instead of targets.*
Low signal:	Insufficient incubation with primary antibody.	*Incubate with primary antibody for longer periods or at warmer temperatures. Ideally, this would be identified during antibody titrations.*
	Insufficient secondary antibody.	*Increase secondary antibody titer. Should be identified during titrations. May indicate secondary antibody degradation.*
	Excessive blocking.	*Decrease block time or blocking reagent (from 5% to 1% milk, for example).*
	Transfer was incomplete.	*Check transfer efficiency using Coomassie, Fast Green, or Ponceau solutions.*
	Protein of interest is below detectable levels.	*This should be identified during antibody and protein titrations. Increase amount of protein loaded.*

	ECL prepared poorly.	*Prepare ECL just before use. May be a sign that the solutions have aged or were left out too long.*
	Sodium azide present.	*Only an issue for chemiluminescence; wash blot thoroughly and make sure that solutions are free of sodium azide.*
	Membrane stripped and reprobed too many times.	*Just start another blot.*
Nonspecific bands:	Antibody has some nonspecific binding.	*Adjust antibody titer or get a new antibody; some antibodies just have nonspecific binding.*
	SDS used in the presence of antibodies.	*Omit SDS after immunoassay, as it interferes with antibody binding.*
Diffuse bands:	Too much protein loaded into gel.	*This should be identified during antibody and protein titrations. Decrease amount of protein loaded into gel.*
	Antibody concentration too high.	*This should be identified during antibody and protein titrations. Decrease antibody titer.*
Blank areas:	Incomplete transfer, air bubbles in "sandwich."	*Check transfer efficiency using Coomassie, Fast Green, or Ponceau staining.*

SOLUTION RECIPES

mqH$_2$O stands for ultrapure water. β-mercaptoethanol, β-ME.

10% SDS	
SDS	25 g
mqH$_2$O	250 ml
Wear a mask to avoid SDS inhalation.	

30% acrylamide	
99.9% acrylamide	58.4 g
Bis N,N'-methylenebisacrylamide	1.6 g
mqH$_2$O	200 ml
Neurotoxic—wear a mask to avoid inhalation. Store at 4°C in a dark bottle.	

1.5M Tris pH 8.8	
Tris base	36.6 g
mqH$_2$O	200 ml
Use a special Tris electrode.	

1M Tris pH 6.8	
Tris–HCl	3 g
mqH$_2$O	50 ml
Use a special Tris electrode.	

10% ammonium persulfate (APS)	
APS	1 g
mqH$_2$O	10 ml
Aliquot and store at –20°C until use; store at 4°C for up to 3 weeks. Use 2.5 g for 25%.	

TBST (0.1%)	
TBS	250 ml
Tween-20	250 µl

Phosphate-buffered saline (PBS)	
NaCl	32 g
KCl	0.8 g
Na$_2$HPO$_4$	5.76 g
KH$_2$PO$_4$	0.96 g
mqH$_2$O	To 4 L

PBST (0.2%)	
PBS	250 ml
Tween-20	500 µl
0.1% for PVDF membranes.	

Blocking buffer (4% blocker)	
Blocker	4 g *4 ml if liquid*
TBS/PBS	100 ml
Adjust weight and volumes to different blocking strengths. I use this as a start.	

2× SDS-PAGE Loading Buffer	
1M Tris pH 6.8	3.76 ml
10% SDS	1.2 ml
Glycerol	6 ml
β-ME	1.5 ml
1% Bromophenol blue	2 ml
mqH$_2$O	4.74 ml

Filter, aliquot, and store at –20°C. Add urea to 4–8 M if necessary. If urea is included, urea must be present during electrophoresis to keep samples denatured.

Optimized homemade ECL
100 mM Tris–HCl, pH 8.8
1.25 mM luminol
2 mM 4-iodophenylboronic acid
5.3 mM hydrogen peroxide
From Haan and Behrmann[21]. Mix fresh from stock solutions for lowest background. Keep luminol and peroxide in the dark. Use if you don't want to order commercial ECL.

Blotting buffer (aka transfer buffer)	
Tris base	12.11 g
Glycine	57.6 g
10% SDS	40 ml
Methanol	800 ml
mqH$_2$O	To 4 L

Tank buffer (aka running buffer)	
Tris base	12.11 g
Glycine	57.6 g
10% SDS	40 ml
mqH$_2$O	To 4 L

Fast Green stock		Fast Green destain	
Fast Green	0.11 g	Methanol	50 ml
Methanol	50 ml	Acetic acid	10 ml
Acetic acid	10 ml	mqH$_2$O	50 ml
mqH$_2$O	50 ml		
Can pour back into bottle and reuse solution.			

Coomassie Blue stain		Coomassie Blue destain	
Coomassie Blue G-250	0.5 g	Methanol	80 ml
Methanol	50 ml	Acetic acid	20 ml
Acetic acid	10 ml	mqH$_2$O	To 200 ml
mqH$_2$O	To 100 ml		

Ponceau Red S stock		Acrylamide gels
Ponceau S	0.5 g	*The solutions vary depending on your gel casters and your percentage of acrylamide. Increase percentage to see smaller proteins, decrease to see larger proteins. Get recipes from someone who uses the same equipment as you. Use 25% APS for minigels, 10% for larger gels. Order TEMED from a supplier.*
Acetic acid	5 ml	
mqH$_2$O	To 500 ml	

PVDF stripping buffer		PVDF strip wash buffer	
Guanidine HCl	573.18 g	TBS	250 ml
Tris–HCl	3.152 g	Nonidet P-40	125 µl
Nonidet P-40	2 ml		
mqH$_2$O	To 1 L		

pH to 7.5. Add β-ME to 0.1M to an aliquot of the solution before use.
You'll need to add heat to eventually make the solution clear up (mix completely). It's endothermic and, without heat, your solution won't mix properly and will always be cloudy. The solution works even when cloudy, but your PVDF and proteins will likely last longer in properly mixed solution.

Alternative stripping buffer		Tris-buffered saline (TBS)	
Tris base	0.76 g	Tris–HCl	28.08 g
SDS	2 g	Tris base	2.68 g
β-mercaptoethanol	700 µl	NaCl	35.08 g
mqH$_2$O	To 100 ml	CaCl$_2$	1.04 g
pH to 6.8 using HCl.		mqH$_2$O	To 4 L

CHAPTER 8

Immunoprecipitation

IMMUNOPRECIPITATION NOTES

Immunoprecipitation (IP) is used to separate proteins that are bound to a specific antibody from the rest of a sample, while co-IP is used to identify protein–protein interactions between the protein that bound to the antibody used for IP and additional proteins that are detected by immunoblotting. It works by binding antibodies to Protein A, Protein G, or a lab-created mix of the two called Protein A/G. Proteins A and G are bacterial proteins that bind very well to antibodies. Proteins A and G bind differently to antibodies from different species, so you can look up a table to identify which protein will work better for you, or you can just order the Protein A/G (it's not much more expensive, if at all). These proteins are bound to agarose beads (or patented variants) to give them weight. The basic idea is that you incubate your IP antibody with the agarose beads conjugated to Protein A/G (some labs call these beads "bugs") and then incubate your sample protein solution with the antibody–agarose bead complex. You then centrifuge the sample at low speed, and the beads–antibody–bound protein complex will form a pellet, which you can separate from the supernatant and resuspend in another solution. Congratulations, you have performed an IP. If you want to perform co-IP, you take your IP product (collecting proteins in a very gentle lysis buffer to preserve protein complexes), run in SDS-PAGE, and perform immunoblots with antibodies that target proteins you think interact with your IP product protein(s). IP controls are similar to immunoblot controls, except that it really helps to have two antibodies that share the same target, so that you can immmunoprecipitate with one, and probe an immunoblot with both (if the IP was true, then both antibodies will stain the same bands). Again, the Abcam Protocol book is a great resource, as is the Thermo Scientific Tech Tip #64. Note that co-IP is different from Far-Western blotting, in that Far-Western blotting is examining protein–protein interactions (like co-IP) but uses a labeled "bait" protein to pull down interacting proteins, instead of using antibodies.

Basic Molecular Protocols in Neuroscience: Tips, Tricks, and Pitfalls. DOI: http://dx.doi.org/10.1016/B978-0-12-801461-5.00008-3

One should remember that IP, immunoblotting, and immunohistochemistry are considered three different techniques for a reason. IP has different experimental conditions than immunoblotting, both of these are different from immunohistochemistry, and those three are different from ELISAs (not covered in this book); basically, some antibodies work for one or more applications, and an antibody that works for all applications is exceedingly rare (I'm not sure one exists!). Further, IPs depend on both the ability of your primary antibody to bind the correct protein and the antibody's ability to bind Protein A/G, so if one of those two things doesn't happen, you won't get data.

IP PROTOCOL

1. You can preclear the protein lysate by adding Protein A/G-agarose beads to the lysate for 10−30 min. Centrifuge at 1000g to collect beads; transfer supernatant to new tube.

   Preclearing is reported by some groups to have big effects. Others report no effects. I suspect this is mostly up to personal tastes.

2. Add 100−500 µg of total protein to a new tube. Add 0.2−2 µg of primary antibody to this tube. Incubate for 1 h to overnight at 4°C.

   I would start with lower amounts of protein and antibody and perform titrations. I used Coomassie Blue staining of the acrylamide gel in order to determine how much antibody−protein I needed to get a nice band. This can also identify (as it did for me) if your antibodies even work for IP. Remember that the Protein A/G typically runs in gels at 50 kDa, so I would not use IP for 50 kDa proteins, because you will have a band there in any case.

3. Add Protein A/G-agarose beads to protein−antibody mixture. Incubate for 1 h at 4°C with agitation.

   There are claims[22] that more protein is collected by adding the antibody to the Protein A/G beads before ever introducing them to the protein sample; I haven't tested this.

4. Centrifuge at 1000g to collect beads. Discard supernatant.

   I would save the supernatant and use that as another kind of control in the ensuing immunoblot. If there is still immunoreactivity in the supernatant, then your IP was incomplete or the antibody isn't so specific (or you didn't put in enough antibody to bind all the copies of your protein of interest).

5. Wash beads four times in **RIPA** buffer or **PBS** (I used **PBS**; see the Immunoblotting Solutions section in Chapter 7). After washes, resuspend sample in SDS-PAGE loading buffer and use at your leisure.

One group claimed to get less background from their IPs if they conjugated HRP to Protein A/G[23]; my IPs didn't have high background, so this wasn't necessary. Some groups argue that IP samples are more stable if dry and break down in SDS-PAGE loading buffer after a week at −20 °C; they say that it is more stable at −80°C. You can test yourself if this is the case.

IP CONTROLS

1. *Lysate control*: run IP sample and IP supernatant together to identify if IP has occurred.
2. *Immunoblot with second antibody control*: using the same antibody for IP and for identifying the protein via immunoblot can have multiple issues. First, the excess of heavy and light chains for the same antibody can give a false positive or high background for the secondary antibody (if your second antibody is from a different host, both issues are removed). Second, if your IP antibody has some spurious binding, using the first antibody again won't separate it from specific IP. Using a second antibody against the same protein has a chance of isolating your protein of interest in the IP.
3. *Peptide preincubation control*: still popular, but still has issues. If you get a band, the antibody may have a higher affinity for the free protein than for the free antigen, or you may have spurious binding. If you don't get a band, either the antibody is specific or has a higher affinity for the free antigen than for whatever protein is happens to bind. Combine with other controls.
4. *Positive and negative controls*: still very helpful to determine if protein of interest is present.

IP TROUBLESHOOTING

Issue	Possible Cause	Possible Solution
High background:	Carryover of proteins that are not detergent soluble.	*Remove supernatant immediately after centrifugation so that detergent-insoluble proteins remain in the pellet.*
	Incomplete washing.	*Increase number of washes.*
	Nonspecific binding to beads.	*Preclear lysate with beads. If nonspecific binding is to Protein A/G, preload antibody to beads, and store in blocking solution.*
	Antibody is binding spuriously.	*Use affinity-purified antibodies to decrease background. Use positive and negative controls to test antibody specificity.*
	Too much antibody is being used.	*This should be identified during antibody titrations. Decrease antibody.*
	Too much protein in lysate.	*This should be identified during antibody titrations. Decrease protein amount.*
	Antigen degrades during IP.	*Include protease inhibitors in lysis buffer and possibly in other buffers.*
	Aggregated proteins in lysate.	*Centrifuge at 100,000g for 30 min to remove aggregated proteins.*
	Bands are actually antibody chains.	*IgG heavy chains appear at ∼55 kDa, light chains appear at ∼28 kDa.*
High amount of antibody eluting:	Too much antibody eluting with protein.	*Reduce amount of antibody. Use a gentle glycine buffer to reduce antibody elution.*
No eluted target protein detected:	Target protein isn't present.	*Use positive and negative controls to determine if this is the case. Antigen abundance may be low.*
	Insufficient antibody for IP.	*Increase amount of antibody. This should be identified during antibody titrations.*
	Target protein has not eluted from beads.	*Determine if your elution buffer has the proper pH or try a new elution buffer.*
	Antibody has not bound to beads.	*Determine if you are using the correct bead type (Protein A or Protein G).*
	Beads broke down.	*Use fresh beads or change supplier.*
	Insufficient incubation time.	*Increase incubation time of antibody with protein, antibody with beads, or both.*
	Interfering substances (DDT, 2-ME, urea) in sample.	*Antibody may not detect denatured protein; if so, change lysis or sample buffer.*

SOLUTION RECIPES

IP PBS

NaCl	0.15 M
Sodium phosphate	0.01 M
pH to 7.2.	

RIPA Buffer

NP-40 or Triton X-100	1%
Sodium deoxycholate	1%
SDS	0.1%
NaCl	0.15 M
Sodium phosphate	0.01 M
Aprotinin	1%
EDTA	2 mM
pH to 7.2.	
Can use 50 mM Tris–HCl.	

Elution Buffer

Glycine HCl	0.1–0.2 M
pH to 2.5–3.0.	
Alter for different elution strengths.	

NP40 Buffer

NP-40	1%
NaCl	0.15 M
Sodium phosphate	0.01 M
Aprotinin	1%
pH to 7.2.	
Can use Triton X-100.	
Can use 50 mM Tris–HCl instead of phosphate.	
Add EDTA to 2 mM.	

TN Buffer

NaCl	0.15 M
Tris–HCl	0.05 M
pH to 7.2.	

IP Wash Buffer

Tris base	0.025 M
NaCl	0.15 M
EDTA	0.001 M
NP-40	1%
Glycerol	5%
pH to 7.4.	

CHAPTER 9

Perfusion and Immersion Fixation

PERFUSION FIXATION NOTES

Perfusion fixation is a method of tissue fixation for histology experiments. Tissue fixation is using a chemical or mix of chemicals to preserve tissue for histology. This chapter describes transcardial perfusion fixation, in which fixative is pumped through the heart and the brain (and the rest of the body, unless you prevent this) and fixes the tissue through the outside and inside (ventricles) of the brain. Perfusion fixation gives superior fixation to immersion fixation, wherein a piece of tissue is dissected out and put into fixative solution as quickly as possible. Immersion fixation fixes from the outside and leaves the insides rotting until the fixative gets there. For very tiny samples (skin biopsies and the like), immersion fixation works fine (and transcardial perfusion of a patient is, to say the least, unethical).

For experiments involving tissue fixed with formalin and stored in paraffin wax, researchers use so-called antigen retrieval steps. I have never needed to perform antigen retrieval for any of the antibodies I've tested, using the methods outlined here. These antigen retrieval steps typically use heat and sometimes pH to reduce fixation-induced cross-linking and therefore expose the antibody's preferred binding site(s). These steps typically require a pressure cooker or similar apparatus. If this is necessary for your experiments, most antibody retailers have protocols on their web sites, as does www.ihcworld.com.

To plan your immunohistochemistry (IHC) or *in situ* hybridization (ISH) experiments, you need fixed tissue (you can use fresh tissue, but you'll need to fix it along the way) and to answer a couple of questions: What should I use to fix the tissue? When I need to get slices, how will I get them and how thick will I cut them? How do I store my fixed, sliced tissue? Let's look at each of these questions in turn.

Tissue Fixation

What chemicals should you use to fix the tissue? My main experience is with paraformaldehyde (PFA) and acrolein. There is also glutaraldehyde,

Basic Molecular Protocols in Neuroscience: Tips, Tricks, and Pitfalls. DOI: http://dx.doi.org/10.1016/B978-0-12-801461-5.00009-5

Bouin's solution (a mixture of picric acid and formaldehyde), picric acid, formaldehyde (aka formalin), and probably others I don't know about. I have avoided Bouin's and picric acid in general because, once picric acid sits for a long time, it forms crystals that explode upon any kind of mechanical shearing—that is, open an old bottle up and lose a floor of the building (no, really, they call in the bomb squad to deal with it). It may give superior fixation, and I'm not going to use it. Glutaraldehyde and acrolein fix in a similar manner. PFA has to be handled gently, as you need a hot solution (around 50°C) to get PFA into solution, but if you raise the temperature too high (58°C, I think), it degrades into formaldehyde and formic acid and fixes a little differently, usually forming an acidic solution that can tear up your tissue.

For most people, 4% PFA works just fine. However, you need to keep your targets in mind. For instance, you can stain for serotonin in tissue slices, but your antibody recognizes serotonin that reacted with glutaraldehyde; if you don't fix with glutaraldehyde, you won't see any staining because the serotonin in your sample is not what the antibody is looking for.

In my experience, a mixture of PFA and acrolein gives superior fixation, producing tougher tissue and better structural integrity, but PFA fixation was still acceptable. Acrolein is nasty stuff and you should only use it under a working fume hood, or you risk burning any exposed membranes. Exposed membranes include your eyes, nose, mouth, throat, and lungs. Use any and all fixatives in a fume hood.

If you plan to perform ISH experiments, 4% PFA deactivates most RNases, but acrolein also deactivates them. I perfuse animals for RNA-based experiments with PFA and acrolein and keep all my other solutions and tools RNase-free. While RNA should be fixed in the tissue, similarly to the proteins, keeping your slices RNase-free is good for peace of mind and allows you to use RNA probes for ISH experiments (if necessary).

I typically fix the tissue, leave it in sucrose solution until it sinks, cut the tissue, and leave it in cryoprotectant solution (see Recipes section); this gives me maximum flexibility for performing experiments. I don't like to prepare and cut the tissue and perform IHC or ISH in the same day, as this limits your flexibility. Other cryoprotectant solutions can be used (such as sodium azide in buffer), but do not work as well as the cryoprotectant solution described here. Use these other solutions if ingredients in the cryoprotectant would compromise your experiment.

Tissue Slicing and Storage

What thickness should you slice tissue at to perform IHC or ISH? That really depends on the experiment and how you plan on performing the experiment, physically: on slides, free floating in wells with a mesh, or free floating in ceramic wells? Each has advantages and disadvantages.

Slices on slides can be very thin (6 μm-thick slices are common), and you get very nice resolution because you only have one layer of cells present (and so nuclear staining by 4',6-diamidino-2-phenylindole (DAPI) or propidium iodide (PI) only shows that one layer). However, IHC or ISH on slides only has two dimensions that the antibodies or probes can bind in, and so in practice uses more antibody or probe to get equivalent staining (though cutting thinner slices can ameliorate this somewhat). Additionally, you usually have to cut these thin sections and put them immediately on slides, as they can fall apart elsewhere. Storing these thin sections on slides precludes the use of cryoprotectant, the slides need to be stored at −80°C, and the tissues are degrading over time. You will need to use a refrigerated slicing unit capable of cutting at these thin ranges (commonly, this machine is called a cryostat).

Free-floating reactions use much less antibody or probe, but you need thicker slices to avoid tearing them apart during your washes. I've used a mesh that the tissue lays on, which makes washes easy (take mesh out of wells, place in new wells. Ta-da! A wash). However, the tissue needs to be thicker so that the mesh doesn't destroy it. In my experience, slices need to be at least 50 μm thick to survive the mesh, which can leave at least three layers of cells (and thus complicates DAPI or PI staining). To use a plate without any mesh, you use a paintbrush and physically move each slice between washes. This is gentler on the slices than the mesh, at the expense of time, as you have to move every single slice, gently, for every single wash, and you want multiple slices per well to have technical replicates for the experiment. Cutting at this thicker ranges precludes the use of most cryostats but allows the use of microtomes. I prefer freezing-stage microtomes with a separate refrigeration unit—stages that require continuous addition of dry ice tend to thaw at inopportune times and require you to focus on both your slices and the stage. You can store these slices on slides at $-80°C$ or in cryoprotectant solution.

IMMERSION FIXATION PROTOCOL

This protocol describes fixation of isolated pieces of tissue or cells in culture. It is very simple and easy to use but the results are not nearly as good as perfusion fixation.

1. Set up surgical tools, storage for fixed tissue, fixative solution, and saline solutions. Keep fixative solution on ice.

Cell Culture (Grown on Coverslips)

2. Aspirate culture medium. Rinse $3\times$ with PBS or your chosen wash solution.

> If your cells cannot handle the fixative solution without being destroyed, you can mix fixative with culture medium to "prefix" the cells. The exact amount of fixative for this can be empirically determined. Start with a half-strength solution (i.e., full strength is 4% PFA, mix 50/50 with medium to leave a 2% PFA solution).

3. Cover cells with ice-cold fixative solution. Let sit for $15-20$ min at room temperature.
4. Rinse $3\times$ with PBS.
5. Begin IHC or ISH protocol.

> I have not found a good way to store cell culture coverslips. If you need to store them, store fixed samples in a 4°C refrigerator.

Extracted Tissue

6. Begin dissection procedure. Put dissected tissue into ice-cold fixative solution.
7. Put tissue/fixative solution into 4°C overnight.
8. Remove fixative solution. Immerse tissue in 30% sucrose solution until tissue sinks (about 2 days).

> If you do not put the tissue into sucrose solution, the freezing process that allows you to cut slices will transform your tissue into a holed mess and waste your time performing this procedure in the first place.

9. Begin tissue slicing procedure.

PERFUSION FIXATION PROTOCOL

This protocol describes perfusing rats. Perfusing mice is very similar, but the small size of the mouse means that anticoagulant isn't necessary and you use far less solution. This protocol should work for most rodents. While the principles remain the same, if you seek to perfuse larger animals or nonmammalian model species, you should seek out a more specialized text.

1. Set up surgical tools, storage for fixed brain tissue, fixative solution, saline solution, and pumps. Test pumps before use to ascertain which parts work and which need to be fixed or replaced (it's really no fun when you've gone through all this trouble and you perfuse an animal with ... air). Start moving saline solution through the pump system, so that when you put the needle into the animal, there is no air moving through the animal but solution instead.

   I opt for perfusion fixation instead of immersion fixation due to how complete the fixation is; immersion fixation works from the outside in, and whole tissue is rotting on the inside while the fixative is slowly working its way there. Perfusion exposes the outside and the inside (via the blood vessels) of the tissue and preserves better. Immersion fixation works for free tissue slices and cells in culture, because a monolayer or bilayer of cells can't keep out the fixative.

2. Keep fixative solution on ice. Add anticoagulant (heparin, for example) to saline solution.

   Again, anticoagulant seems necessary for rats but not mice, due to size.

3. Anesthetize animal, and wait until animal falls unconscious. Confirm that animal is unconscious by pinching the animal's toe. If the animal responds, do not proceed until animal is anesthetized enough that it does not respond to the toe pinch.

   I use a mixture of 90% ketamine and 10% xylazine to anesthetize the animal; use whatever amount your Animal Care and Use Committee (ACUC) deems necessary (due to differences with ACUCs, I can't recommend a specific amount of each anesthetic, but 70–100 mg ketamine/kg of body weight + 5–10 mg xylazine/kg of body weight works). The animal should take a minute or so to fall unconscious; if the animal collapses immediately, you either used too much or got it directly into an artery and you have to move fast, because the animal will be dead in another minute and you won't get a good perfusion that way.

4. Move animal to perfusion space. Open thoracic cavity and cut ribs until you have exposed the heart. It should still be beating.

5. Insert needle into left ventricle of the heart, making sure that it stays in the site and doesn't fall out by itself.

   Try to make sure that the needle won't move and that it's not in your way; few things in this protocol are more frustrating than having a great start that ends in failure because you knocked the needle out of the animal.

6. Cut right atrium. If you want to limit the amount of fixative you use, clamp the descending aorta to avoid whole-body perfusion and limit fixative to the brain.

   This is an issue with the rat, where an adult can take up to 400 ml of fixative. Mice take up much, much less fixative than rats.

7. Allow saline solution with anticoagulant to perfuse through animal.

   The timing and volume of saline and fixative perfusions is up to you, but keep them consistent for your experiment (I use 3 min for young rats at a constant flow rate). This step prevents coagulated blood from preventing full fixation. You can simply perfuse the animal with fixative and omit this step, but some tissues may not get as much fixative as others due to clots blocking fixative flow.

8. Switch perfusion from saline solution to fixative.

   I typically perfuse the animal with fixative for 12 min. Adjust the time for the size of the animal and your specific flow rate. Mice require lower flow rates than adult rats, for example. Flow rates generally depend on your selected pump or other means of generating fixative flow. Remember, the point is to have each animal in your experiment receive equivalent volumes of fixative. I keep my pump pressure constant and therefore use time as a proxy measurement for how much fixative each animal receives.

9. Animal's body and extremities should be pale and hard, unless you clamped the descending artery. Remove brain from skull (or other tissue of choice) and place into a tube with additional fixative.

10. Leave brain in fixative; this is called the postfix. Postfix for a number of hours at 4°C.

    The amount of time you leave a brain in postfix is important. I left my brains in for 2–5 h for the sake of consistency, but you want to get a good system going. If you under-fix, your tissue won't hold together well and won't stain well. If you over-fix your tissue, the antibodies won't be able to access their targets and the tissue won't stain well. Either use an established time or test a couple of fix times, using a known good antibody for your IHC.

11. When postfix is finished, remove fixative and leave brain in 30% sucrose solution at 4°C until the brain sinks.

You might be thinking, why sucrose? For the slicing techniques I employ, the tissue is frozen on a stage. Sucrose pushes water molecules out of the way, so that when tissue is frozen, ice crystals don't form within the tissue. If you get ice crystals in the tissue, they will physically tear apart your tissue and it will be full of holes. I use the sucrose solution and don't encounter problems. Some protocols start at 10% sucrose and follow with 20% and then 30% to get the brain to sink. I start at 30% sucrose and leave the tissue alone for at least 3 days. After 3 days, the brain has sunk and I don't get any holes in my tissue. Alternatively, there are commercial freezing solutions that can perform the same task as the sucrose and be put into −80°C.

SAMPLE STORAGE

1. When brain has sunk in sucrose solution, remove brain from solution and immediately begin slicing procedure.

You can leave tissue in the sucrose solution for an unspecified period of time; some people leave it for months. The issue is mold. Sucrose is mold food. Living in any subtropical area lets mold breed freely. Mold will destroy your tissue and your ability to stain tissue. As such, I like to get my tissue into cryoprotectant and into the cold (to prevent mold from surviving to ruin my work) as soon as possible. You can also place the brain in a gelatin matrix to keep smaller tissues (like olfactory bulbs) from falling apart, but I haven't tried it, as the gelatin solution includes egg yolks. Any contamination of your gelatin solution by egg whites, which are full of avidin, will give false negatives when using the ABC kit.

2. Using a freezing-stage microtome, place brain onto freezing stage and cover with embedding compound.

There are different microtomes and cryostats that you can use, but I prefer to use microtomes with stages maintained by an attached refrigerator unit for consistency. Needing to add dry ice continuously, in order to keep the tissue frozen, is at best a nuisance and at worst going to cost you tissue integrity (when you forget). Cryostats are used for much smaller slices than I like to cut; I prefer free-floating IHC, which is impossible (or next to it) with the 2–20 μm-thick slices from the cryostat.

3. When the embedding compound freezes, you can begin cutting tissue slices at your preferred thickness.
4. Transfer tissue slices from the microtome blade to cryoprotectant solution using a paintbrush.

I typically store tissues in 6-well or 12-well plates with each well half filled (adjust to your liking) with cryoprotectant. When the wells are empty of tissue, the tissue will unfurl itself from your paintbrush into the solution. You will eventually need to (very gently) shake or spin the paintbrush to get the tissue off the bristles. Do not press the paintbrush against the bottom, as this will cause the bristles to spread out and likely destroy your tissue slice.

5. When finished, store tissues in cryoprotectant solution at −20°C.

Tissue slices keep indefinitely; I have had no problems with mold or rotting tissue for years (some researchers have tested tissue that was in cryoprotectant for at least 5 years with no loss of staining quality). I have not encountered problems with cryoprotectant evaporation, either. Different reports indicate that you should leave your tissue in cryoprotectant for 5 days for best results in IHC or ISH staining.

SOLUTION RECIPES

4% PFA

PFA	40 g
dH$_2$O	500 ml

Heat water to 50°C, add PFA. Add NaOH to clear solution. Cool to RT.

5× PBS	200 ml
dH$_2$O	To 1 L

Filter, pH to 6.8.

0.9% saline

NaCl	9 g
dH$_2$O	To 1 L

Optionally, add DEPC and autoclave.

4% PFA/2.5% acrolein

Make PFA solution as described above, but only add water to 975 ml. Add 25 ml acrolein just before use.

1× PBS

5× PBS	200 ml
dH$_2$O	To 1 L

30% sucrose

Sucrose	150 g
dH$_2$O	To 500 ml

Autoclave. Add 1 ml of DEPC before autoclaving if RNase-free solution is desired.

Antifreeze cryoprotectant

5× PBS	200 ml
Sucrose	300 g
Ethylene glycol	300 ml
PVP-40	10 g
dH$_2$O	To 1 L

Optionally, one can add DEPC and autoclave the PBS and sucrose solution, then add the other ingredients for RNase-free solution. Keep tissue for a minimum of 5 days in solution for better staining.

5× PBS

Na$_2$HPO$_4$	9.2 g
NaH$_2$PO$_4$	2.35 g
NaCl	40.9 g
dH$_2$O	To 1 L

CHAPTER 10

Immunohistochemistry

IMMUNOHISTOCHEMISTRY NOTES

To start, a great resource for immunohistochemistry (IHC) is Hoffman et al.[24] I highly recommend that you read it if you are interested in learning more about IHC. The protocols that I describe in this chapter are adaptations of the protocols in that paper.

So what is IHC? It is the process of using antibodies to bind a protein in a tissue section and use some chemistry to visualize where that antibody bound (and therefore localize the protein). Immunocytochemistry (ICC) is almost the same thing as IHC; the steps are the same after fixation. ICC examines cell culture preparations and IHC examines tissue sections. Only sticklers like me insist on using the specific terms for the different kinds; most people will just call both IHC and ICC by the same acronym.

Antibody binding in tissue works a little differently from antibody binding in immunoprecipitation or immunoblotting. In those techniques, antibodies bind to proteins that are free in solution or fixed to a membrane. IHC binds proteins that are fixed into some three-dimensional conformation, whether by your tissue fixation solution or by being in their actual site of origin (like receptor proteins are fixed in the membrane). As such, there is a difference in both the number of binding sites and the types of binding sites available. An antibody that works in IHC may or may not work in immunoblotting or immunoprecipitation.

There are many different issues to consider when planning your IHC experiment; we'll discuss each in turn.

Types of IHC Staining

There are currently two main branches of IHC staining: colorimetric and fluorescent. Colorimetric IHC, much as in immunoblotting, forms a colored precipitate at the site of antibody binding to protein. Unlike immunoblotting, however, colorimetric staining can be permanent, as the reaction product is held in place (which can be easily washed off of

Basic Molecular Protocols in Neuroscience: Tips, Tricks, and Pitfalls. DOI: http://dx.doi.org/10.1016/B978-0-12-801461-5.00010-1

blotting membranes); permanence depends on the stain of choice. Typically, colorimetric staining is used for qualitative experiments ("Where's my protein?"). While some groups claim to pursue quantitative colorimetric staining, I don't buy it: colorimetric staining works by enzymatic action and therefore the reaction continues until there are no more reagents or there is a loss of physical space to deposit more reactant, and either condition can make it appear to have more or less protein.

Many colors are available for colorimetric staining (Vector Labs sells a number of good reagents). I typically perform colorimetric staining with 3,3-diaminobenzidine (DAB) with nickel salts (Ni-DAB). DAB, by itself, stains the entire tissue slice a light brown, while a darker brown indicates the presence of your protein. Ni-DAB stains black and white and is easier on my eyes to see the difference between staining and not staining (black and white tissue slices are easier to examine than gradients of brown); additionally, Ni-DAB stains slightly blue and brown if your titration isn't optimal, and this really helps to decide how much antibody to use. The other benefit of DAB (nickel or not) is that the product is permanent, while some other colorimetric stains are not permanent. Colorimetric stains may need a counterstain (like cresyl violet or methyl green) in order to see structures appropriately.

Fluorescent IHC is when your antibody (primary, secondary, tertiary, or streptavidin, or whatever you use to bring in your signal molecule) is conjugated to a fluorescent molecule. Fluorescence is, in my opinion, preferable for quantitation and multiple labeling experiments in IHC. However, while fluorescent IHC reactions are shorter overall and easier to perform, there is a big cost: the fluorescence is dying out as soon as you're done with the experiment. Even with antifade reagents, after a month of being kept in the dark and in the refrigerator, fluorescence is lost from slides, gone, repeat the experiment. As such, it's usually best to check your fluorescent IHC either the day you finish the experiment or within a week of doing so; if you choose to count after a couple of days (some report that this lowers background noise), then choose a consistent timeframe for examination so that any quantifiable changes are not due to differences in fluorescent quenching. The two available fluorescent counter stains are 4',6-diamidino-2-phenylindole (DAPI) or propidium iodide (PI). Both of these bind to DNA and therefore stain nuclei: DAPI stains blue and PI stains red/orange. If you're using a red/orange fluorophore, then don't use PI. The resolution of these counter stains depends on the thickness of your slice; slices as thick as 50 μm can hold three or more layers of cells, and therefore complicate analysis of the counterstain.

Secondary Antibodies and Blocking Buffers

As in immunoblotting, we use blocking reagents to block nonspecific antibody binding in the tissue, so that the staining we observe is just due to our primary antibody. Typically, BSA, casein, and normal serum are used as blockers. There are commercial blockers available, but I don't like to use them because I don't know what's in them. If they happen to include BSA, then I can use goat antibodies all day long and not know why I'm getting such high background signal. If they include rabbit serum, then my rabbit antibodies will look especially terrible and I won't know because the block solution's components are protected information! I use serum to block and try to use secondary antibodies from a single host (at least within a single reaction; see Multiple Labeling section). Avoid using milk as a tissue blocker; it has so many different protein components that stick all over the tissue slice that you get some background and/or ugly slices (including biotin, which will give false positives in an avidin–biotin complex (ABC) reaction, and phosphoproteins, which can give false positives for phosphorylated targets).[25] You can test different blockers for your reaction, but I've had good experiences just using serum from the secondary antibody host animal (e.g., use normal goat serum to block if you're using a goat secondary antibody).

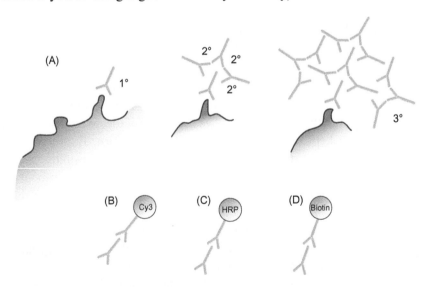

Amplification of the Signal

Amplification of the signal is admittedly more common in IHC than in immunoblotting. So what do I mean by amplification? Let's say that you have a protein of interest that has three binding sites for your primary (first) antibody (all of these numbers are examples). If you directly conjugate some signal molecule to your primary antibody, there will be three signal molecules (max) per copy of your protein of interest. Three signal molecules don't generate a lot of signal. Well, let's add a secondary antibody to the mix, and say that there's four sites for the secondary antibody to bind your primary (there's typically a lot more, but this is an example). If each secondary antibody has a signal molecule conjugated to it, there are now 12 signal molecules (max) per copy of the protein of interest (see figure above). I routinely use a form of amplification called the ABC reaction (comes in a kit; get the unmixed kits, if possible). Avidin is found in egg whites, but streptavidin (from *Escherichia coli*) is a much more stable form of avidin. I use a biotinylated secondary antibody, and each biotin has four binding sites for avidin. Free biotin will bind avidin, and these will continue to bind each other in solution until there is a huge complex of molecules with signal molecules on them (HRP conjugated to streptavidin, in fact). I have also used streptavidin alone with a biotinylated secondary antibody (no free biotin, therefore not an ABC reaction) to amplify my signal; in keeping with the above example, streptavidin amplification would look like this: 3 primary binding sites $\times$ 4 secondary binding sites $\times$ 4 avidin binding sites $= 48$ signal molecules per protein (max). There are also kits incorporating tyramide amplification, which amplifies the signal greatly.

Multiple Labeling

What kind of experiment are you performing? Are you hoping to examine multiple proteins in the same tissue section? While that can be done, there are some additional considerations for that experiment. First, multiple labeling experiments work much better if your primary antibodies are made in different hosts. There are ways to "convert" a primary antibody to another "host" but are far more cumbersome than using antibodies from a different host. For example, one can use goat anti-rabbit Fab fragments to convert a rabbit antibody to a "goat" antibody and use anti-goat antibodies conjugated to a signal molecule to get data. Second, multiple labeling experiments work best if your secondary antibodies come from the same host. As an example: mouse anti-tyrosine hydroxylase and rabbit anti-calbindin primary antibodies are used with goat anti-mouse and goat anti-rabbit secondary antibodies. This combination will stain well together (different hosts for primary antibodies) with low background (one host for the secondaries means that I can block with only goat serum; if secondaries were from different hosts, I would need normal serum (plural *sera*) from each, and the secondaries might cross-react with the different sera). Multiple labeling works best with fluorescent detection; while colorimetric staining can work with multiple labeling, colorimetric staining is forming a precipitate, and one precipitate might crowd out the other (both in terms of not being able to see one precipitate, and physical space for the chemical reaction). There are some who perform multiple labeling with colorimetric systems and have found ways to use the same enzyme (HRP or AP) for both stains,[26] since HRP and AP have different sensitivities. Additionally, while this may be elementary, double-check that your secondaries (or whatever you're using) are conjugated to different fluorophores that show up on different parts of the visible spectrum (both excitation and emission spectra); any bleed over in light emission will give false positives.

ANTIBODY TITRATION PROTOCOL (NI-DAB)

I begin with this protocol when I acquire a new antibody. This gives me an idea of what titers work for staining and which do not. This protocol (mostly) follows the advice of Hoffman et al.[24]

1. Remove tissue slices from cryoprotectant and place into PBSX.
2. Three washes, 10 min each (3 × 10′), in PBSX to wash out cryoprotectant solution.
3. 15′ wash in 1% hydrogen peroxide (most fixatives) or 1% sodium borohydride (if acrolein or glutaraldehyde was present in the fixative).

> *In my experience, this is an important step. We wash in peroxide/NaBH₄ to decrease background noise in the ABC reaction. Ni-DAB precipitates are formed by the reaction of HRP (from the kit) with peroxide. Red blood cells also express a peroxidase, and applying peroxide this early exhausts the ability of that peroxidase to catalyze reactions. The borohydride is used because acrolein forms free aldehydes as part of its fixation process; these aldehydes can bind antibodies and give false positives. Borohydride reacts and frees aldehydes from solution in addition to reacting with peroxidase. You'll see bubbling either way. On another note, 30% hydrogen peroxide is not as stable at 3% hydrogen peroxide. 3% is the bottle you get from any pharmacy for less than $2 USD. I buy the 3% and don't have to worry about degradation of 30% peroxide, which has caused me to get false negatives because the Ni-DAB precipitate never formed.*

4. $3 \times 10'$ washes in **PBS**.
5. 60 min in blocking solution.

> *As described above, I prefer to use serum for my blocking solution. I have noticed absolutely no difference in blocking quality between 1% and 10% serum solutions for most antibodies. Most protocols recommend 10%; I only use 1% serum unless it becomes obvious that I need to do otherwise.*

6. Put tissue into primary antibody solutions. I typically run a logarithmic curve of antibody dilutions: 1:1000, 1:3000, 1:10,000, 1:30,000, 1:100,000, and 1:300,000.

> *While it may be tempting to avoid performing the whole curve, my experience has been that some antibodies work at much lower titers than many researchers expect. A goat anti-calretinin antibody from Chemicon that I've used didn't even show staining until the 1:100,000 dilution. Unless under extreme duress, just test the whole curve. This step is part of why I'm skeptical about the titers reported by different companies.*

7. Leave tissue in primary antibody solution for 48 h at 4 °C.
8. After the 48 h incubation, $3 \times 10'$ in **PBSX**. Save antibody, if desired.
9. 2 h in secondary antibody solution.

> *Most people use secondary antibodies at concentrations like 1:200 or 1:600. I think both of these are a waste of secondary. I use 1:1000 and have seen no difference in staining between 1:500 and 1:1000. I could probably use less, but secondary antibodies are comparatively cheap.*

10. $3 \times 10'$ **PBSX**.

> *Because most ABC solutions require 30 min to incubate, I prepare the solution just before I start these washes.*

11. $60'$ in **ABC** solution. I use much less solution than the company recommends; see Solutions and Notes.

> *The company says to use drops; some protocols say to use a 1:100 dilution of A and B solutions. I use 45 µl of A and 45 µl of B per 10 ml of total solution; much less than recommended to use, but my picture quality is great every time.*

12. $2 \times 10'$ in **PBS**.
13. $5'$ in acetate buffer.

14. 20′ in Ni-DAB solution.

> *While most people think 20 min is insane, allowing the reaction to proceed for such a long period of time (including 48 h of primary incubation) allows me to use much lower titers of antibody than otherwise. Most people only want to leave antibodies on tissue overnight; just try 48 h and see if your picture and antibody titer improve.*

15. 5′ in acetate buffer.
16. 2 × 10′ in **PBS**. Mount tissue on slides (make sure they are flat!), allow to air dry overnight. Follow with coverslipping protocol, moving through graded ethanol solutions, then delipidizing tissue using xylenes (or available alternatives). Apply some adhesive and put coverslip on, gently pressing on coverslip to remove bubbles. Allow to dry (hours) and view under light microscopy.

> *Staining should appear black on a white or very light background. Sometimes the DAB itself lightly stains different structures, acting as its own counterstain. If staining is brown or blue-black, that titration is not optimal (typically the titer is too high). Note that positive and negative controls (like omitting the primary antibody) are very important here, because Ni-DAB can generate a lot of extra noise that, if staining is very weak, would be indistinguishable from a weak signal. This noise is present even after filtering the Ni-DAB solution.*

ANTIBODY TITRATION PROTOCOL (IMMUNOFLUORESCENCE)

This follow-up to the Ni-DAB titration follows the advice of Hoffman et al.[24]

1. Remove slices from cryoprotectant and place into PBSX.
2. 3 × 10′ PBS to wash out cryoprotectant.
3. 60′ in blocking buffer.
4. Place tissue into primary antibody at 10×, 30×, and 100× the amount of antibody that you determined was optimal using the Ni-DAB titration. If you want to use a biotinylated secondary with fluorophore-conjugated streptavidin, use 10×, 20×, and 30× the amount of antibody that you determined was optimal using the Ni-DAB titration.

> *This part can require additional titrations, but I have typically found that company datasheets give a much higher titer (more concentrated) than I successfully used in my own IHC. The additional titrations are usually worth it.*

5. Save antibody if desired. $3 \times 10'$ in **PBSX**.

   If using a monoclonal antibody, this is only to reuse the antibody. If you are using a polyclonal antibody, running the antibody through a reaction once is a cheap form of affinity purification: the various antibodies have already bound the nonspecific proteins and parts that they could, and so (in theory) the remaining antibody is specific for your target. You could, alternatively, do a similar purification by incubating your polyclonal antibody in tissue where the target is knocked out.

6. 2 h in secondary antibody solution.

   Some people increase the amount of secondary antibody because they're performing immunofluorescence; I still use 1:1000 and it works just fine. I have seen no difference in staining between 1:500 and 1:1000 dilutions of the secondary antibody.

7. If using streptavidin, wash $3 \times 10'$ in **PBSX**, followed by $60'$ of streptavidin.

   I use the fluorescent streptavidin at 1:1000, just like the secondary antibody.

8. $3 \times 10'$ **PBS**, mount tissue. Allow to air dry (20 min), cover tissue with antifade reagents, and place coverslip on slide. Gently press on coverslip to remove bubbles. Allow to air dry (20 min or overnight) and apply nail polish (or some plastic adhesive) to edges to seal slide. View under fluorescent microscopy within a week.

   Antifade reagents are crucial but not perfect. Even with antifade reagents, my FIHC signal is gone if left in the dark and in the cold for a month. While you can get a counterstain (like DAPI) included in the antifade reagent, it's not necessarily a good idea if you want to pursue confocal microscopy. The excess DAPI in the reagent can overwhelm the microscope's detection ability; use a separate DAPI reaction if you want the counterstain or at least check if the microscope you wish to use can handle the DAPI.

Staining should appear brighter than background fluorescence. Adjust the fluorescent lamp output so that you see immunoreactivity higher than background noise. Use the lowest amount of antibody possible while getting the same amount of immunoreactivity.

IHC PROTOCOL

The IHC protocols are the same as the titration protocols, except at the titer you calculated for your own antibodies.

CRESYL VIOLET COUNTERSTAIN (NISSL STAIN) PROTOCOL[27]

1. For free-floating sections, wash slides briefly in dH_2O to remove any residual salts.
2. Immerse slides through two 3′ changes of 100% ethanol.
3. Immerse slides through two 15′ changes of 100% xylenes (or alternative).
4. 3×10 min in 100% ethanol.
5. Wash in dH_2O briefly.
6. Stain in Cresyl Violet solution for 15 min.
7. Rinse in dH_2O briefly.
8. Wash in 70% ethanol briefly.
9. 2 min in differentiation solution.
10. Two 3′ changes of 100% ethanol.
11. Two 3′ changes of 100% xylenes, coverslip. Allow to dry in fume hood.

Any counterstain should be undertaken with an understanding of the original stain involved. For instance, DAB precipitates are insoluble in water and alcohol but are sensitive to pH. Older xylenes and alcohols can have pH changes that affect DAB deposits, especially if those deposits are on the surfaces of cells. Slices can show excellent Nissl staining while your original staining has disappeared.

IHC CONTROLS

1. *Negative controls*: control tissue (knockout tissue, tissue that doesn't express the protein, cell lines that don't express the protein) can be useful to identify spurious binding of the primary antibody. Remember that knockouts aren't always perfect: virally induced knockouts aren't 100% effective and may still show some staining (if they don't infect every cell), while some knockouts still generate a truncated (nonfunctional) protein that may still be recognized by your antibody.

2. *Positive controls*: control tissue that does express your protein of interest can identify primary antibody problems, including low antibody titers (primary and/or secondary) or blocking issues.

3. *Peptide preincubation control*: while incubating the antibody with the antigenic peptide before exposing the antibody/peptide mixture to tissue (called a peptide preincubation control) has been a popular choice, it's not necessarily a good control. If you don't see staining after a peptide preincubation control, it can mean the following:

   a. The antibody doesn't bind spuriously, as it bound the peptide completely.

   b. The antibody has a higher binding affinity for the free peptide than whatever it did bind in your tissue (which can still be nonspecific!).

   c. You preincubated in the wrong conditions and now the antibody has no activity (to solve this, simply leave both your antibody with peptide and your antibody alone solutions in the same conditions).

   If you see staining after a peptide preincubation control, it can mean the following:

   a. The antibody binds spuriously, as it didn't bind the peptide completely.

   b. The antibody has a higher affinity for the fixed target than the free peptide.

   I consider the preincubation control to be better than nothing, but the results are ambiguous without further testing; combine it with other controls.

4. *Omit primary antibody*: tests if the secondary antibody or other signal-generating components in the reaction bound to your tissue and gave false positives or background noise.

5. *Omit primary and secondary antibodies*: identifies spurious binding from additional signal amplification components (such as endogenous biotin giving a false positive for the ABC reaction). Unnecessary if omitting the primary antibody doesn't lead to any staining, noise, or otherwise.

6. *Single-color controls*: for multiplex fluorescent IHC, this can identify if your antibodies are binding each other instead of the target

proteins, and if your fluorophores are "bleeding over" into each other's channels (and therefore giving you false positives).

7. *Fixation controls*: some antibodies will not find their targets if the tissue is fixed with incompatible fixative. Additionally, some fixations are just not as good as others, and it can be useful to have a "known good" set of fixed tissue to compare to new tissue to determine if the tissue fixation was acceptable.

8. *Multiple antibodies against different parts of the target*: expensive, but useful, as it is in immunoblotting. If you have multiple antibodies that target different parts of the same protein (one targets the C-terminus, and the other an internal portion, for example), check if their staining patterns are the same. If so, that's a good indication of specificity. Combine with other controls.

9. *Colocalization with mRNA*: combine IHC with *in situ* hybridization (ISH) to see if your antibody staining overlaps with the mRNA staining (though each stain may be in a different compartment of the cell, depending on your target). A very convincing control, if difficult to pull off, because optimal ISH conditions may be incompatible with optimal IHC conditions.

10. *Use antibody in immunoblotting to validate antibody*: this control can be useful but has some drawbacks. This control is considered a success if your immunoblot has bands at the correct weight(s). However, immunoblotting and IHC use antibodies to identify targets in very different scenarios: immunoblotting examines a protein fixed to a membrane, while IHC examines a protein fixed within tissue. Many antibodies that work for one technique don't work for the other, which can give a false negative. Further, antibodies that are specific for one technique can have spurious binding when used in the other technique. This control can also take a lot of time, as you would need to perform controls for both techniques in order to truly assess specificity. While this can suggest specificity, it is best combined with other controls.

IHC Troubleshooting

Issue	Possible Cause	Possible Solution
High background:	Nonspecific binding.	*Test antibodies on positive and negative control tissue; increase blocking reagent or get new antibodies. You can also incubate secondaries at 4° C.*
	Fluorophores have similar emission wavelengths.	*Choose fluorophores with nonoverlapping spectra; it'll make the experiment go much more smoothly.*
	Endogenous biotin.	*Only some tissues have this problem; use a biotin/avidin blocking solution before primary antibody incubation.*
	Avidin introduced into system (issue for ABC amplification).	*This only really happens if you study chickens or you put the tissue in a gelatin/ egg yolk mixture and got some egg white contamination. Choose a different amplification technique or start over.*
	Insufficient blocking.	*Increase block time or reagent concentration (1–5% serum, for example).*
	Incorrect antibody titer.	*Perform antibody titration. Some antibodies need high titers to show up, some don't show up until very dilute titers.*
	Primary antibodies bind to each other instead of targets.	*Use single-color/single-antibody controls to determine if this is the case. Don't mix the antibodies again if you know they bind each other.*
Low signal:	Poor or over-fixation.	*Compare newly fixed tissue to known good tissue to determine if this is the case. Start over if that's the case.*
	Excessive blocking.	*Lower block times or reagent concentrations (from 5% to 1% serum, for example).*
	Incorrect antibody titer.	*Perform antibody titration. Some antibodies need high titers to show up, some don't show up until very dilute titers.*
	Not enough signal amplification.	*Not typically an issue when using the ABC kit, but you can add tertiary amplification (streptavidin, tertiary antibodies, and tyramide kits).*

SOLUTION RECIPES

DAB solution (3.3 mg/ml)

DAB	330 mg
Acetate buffer	100 ml

Aliquot and freeze solution at –20°C. Can reuse aliquots after thawing. Avoid exposure to light.

NiSO$_4$ solution (416.7 mg/ml)

NiSO$_4$*6H$_2$O	41.67 g
Acetate buffer	100 ml

Aliquot and freeze solution at –20°C.
Can reuse aliquots after thawing.

AB solution (0.4% Triton X-100)

Triton X-100	800 ml
1× PBS	200 ml

PBSX (0.1% Triton)

5× PBS	200 ml
Triton X-100	1 ml
dH$_2$O	To 1 L

1% serum solution

Normal serum	200 µl
1× PBS	To 20 ml

ABC solution

A	45 µl
B	45 µl
AB	To 10 ml

1% sodium borohydride

NaBH$_4$	0.3 g
1× PBS	To 30 ml

Streptavidin solution

Streptavidin	Varies
AB	Varies

Cresyl violet solution

Cresyl violet acetate	1 g
Acetate buffer	To 500 ml

Acetate buffer

Na acetate	1.45 g
dH$_2$O	100 ml

pH to 7.5–7.6. Make fresh.

1× PBS

5× PBS	200 ml
dH$_2$O	To 1 L

5× PBS

Na$_2$HPO$_4$	9.2 g
NaH$_2$PO$_4$	2.35 g
NaCl	40.9 g
dH$_2$O	To 1 L

Antibody solution

Normal serum	200 µl
AB solution	To 20 ml

Vary the amount of solution and antibody based on your needs.

Ni-DAB solution

NiSO$_4$ solution	2.4 ml
DAB solution	2.4 ml
Acetate buffer	35 ml

Filter Ni-DAB solution in a Whatman filter (1 or 3 works) before use. Add 300 µl of 3% hydrogen peroxide just before use.

1% hydrogen peroxide

3% H$_2$O$_2$	10 ml
1× PBS	To 30 ml

Differentiation solution

Glacial acetic acid	4 drops
95% ethanol	200 ml

CHAPTER 11

In Situ Hybridization

IN SITU HYBRIDIZATION NOTES

In situ hybridization (ISH) is a means of identifying where mRNAs are present in fixed tissue samples; as IHC identified proteins in fixed tissue, ISH identifies mRNA. ISH is performed by designing an antisense probe to your mRNA target, allowing your probe and mRNA to bind, and visualizing where your probe is in the tissue slice. *In situ* is Latin for "in location," and so you could consider ISH and IHC to be *in situ* techniques. ISH and IHC can be run together in the same tissue sample as an elegant control to show that your ISH probe and your IHC antibody are specific, but the two techniques have different requirements and so this control is difficult to pull off. Many online guides to ISH exist, but one good resource is from Dr. Wilcox,[28] and another is from Dr. Tom Houpt on his web site, the MagnetoWiki.

To get an ISH probe, you can design it much the same way as a PCR primer, except you're only interested in the antisense probe. For ISH, you can use cDNA probes, oligonucleotide probes, or RNA probes. RNA probes are reputed to be the best, followed by oligos, followed by cDNA. RNA probes are typically generated by introducing plasmids into bacterial cultures (like *Escherichia coli*) and purifying the probe. They are very sensitive (in every sense of the word) and require everything to be RNase free at every step before probe binding (after binding, the mRNA/probe is considered by RNases to be double stranded and therefore not a target). Oligonucleotide and cDNA probes are not as sensitive but don't come with the "everything must be RNase free and why won't you cater to my every whim" mentality that RNA probes require. Oligonucleotide probes are typically ordered from companies, unless you have a nucleic acid synthesis machine available. cDNA probes are generated from PCR; the PCR product is followed by asymmetric PCR (differing concentrations of forward and reverse primers) to generate a concentrated amount of probe.

Basic Molecular Protocols in Neuroscience: Tips, Tricks, and Pitfalls. DOI: http://dx.doi.org/10.1016/B978-0-12-801461-5.00011-3

Types of ISH Probe Labeling

Once you have your probes, you must decide how to acquire a signal. Radioactive probes are the most sensitive, but nonradioactive methods have been making great strides in recent years. For radioactive probes, you have a couple of choices: ^{35}S, ^{125}I, ^{32}P, and some others. The iodine and phosphorus isotopes are "hotter" and have higher reactivity and shorter half-lives than the sulfur; sulfur seems to be the best to use unless you have a low abundance target, in which case the room is divided—some prefer to use hotter probes, others contend that sulfur is really just the best overall. Nonradioactive probes are currently split between biotinylated probes, digoxigenin (DIG) probes, and tyramide probes. Biotinylated probes are amplified using an ABC kit or streptavidin, similarly to IHC, though you can also use an anti-biotin antibody and a secondary antibody to amplify the signal. Tyramide probes are amplified using, unsurprisingly, a tyramide amplification kit. DIG probes are amplified using anti-DIG antibodies and other antibodies, if necessary. The draw of DIG probes is that DIG itself is found in plants, and so you should have zero background in your animal tissue from the anti-DIG antibodies. Biotin and tyramide are pursued because, while not as sensitive as radiation, you can still get quite a lot of amplification out of those probes. According to Dr. Wilcox, RNA probes are more sensitive than oligonucleotide probes, which are more sensitive than cDNA probes. Additionally, he claims that ^{35}S is more sensitive or equal to ^{33}P, which is more sensitive than ^{32}P, which is more sensitive than biotin and DIG. He also says that frozen tissue works better than paraffinized tissue. Your mileage may vary, but this is a good start.

You can order probes with these signal-generating molecules attached, or you can attach them yourself using different kits. Either way is costly. Be aware that some kits and some companies will attach molecules via a dideoxynucleotide base; this means you can attach only one molecule, and your signal/noise better be awesome for you to get anything for your time. It is better to have a regular nucleotide, so that you can add many signal molecules to your probe. If you use multiple types of signal molecule, you can perform multiplex ISH, using a combination of different probes (biotin, DIG, fluorescein) and antibodies (anti-biotin, anti-DIG, anti-fluorescein) to generate signals (or different fluorescent molecules on the probes, but this doesn't allow as much amplification or long probe storage). As with fluorescent immunoblot or IHC, make sure that your fluorophores emit nonoverlapping wavelengths of light. For one protocol, see Ishii et al.[29]

Visualizing Your ISH

The process of visualizing your data depends on what kind of probe you use. For radioactive probes, you can submit your tissue to liquid emulsion followed by photography, or autoradiography film. Use of the film has the same issues here that it does for immunoblotting; you need a minimum amount of signal to get the silver grains to react, and there is a point where the film can't detect any further changes. For nonradioactive probes, you have many of the same options you have for IHC: fluorescence and colorimetric staining. Colorimetric stains require that you include an enzyme at some point in your experiment, whether conjugated to streptavidin or a secondary antibody (I've not heard of someone attaching HRP or AP directly to their probe, and you won't get much amplification that way). For fluorescence, you can attach a fluorophore to your probe, streptavidin, or secondary antibody. Afterward, you can visualize your colorimetric or fluorescent ISH reactions using the same methods of microscopy that you used for IHC.

ISH PROBE DESIGN FOR OLIGONUCLEOTIDE AND PCR PROBES

Before performing this step, check the literature and online databases to make sure that someone else hasn't developed a working probe already.

1. Using the BLAST algorithm from ensembl.org or NCBI, enter your sequence of interest.
2. Select "Blast against cDNAs" and "Allow some local mismatch." Look at the generated schematic to identify areas of the cDNA that do not also correspond to other parts of the genome.
3. Find a part of the diagram that has only a single bar, click on the Alignment tab, and make sure it's your gene of interest.
4. The number of bases for your probe depends on what kind of probe you want to design. Oligonucleotide probes can range from 18—50 bases, while PCR-generated probes are much larger (hundreds of bases).

5. Submit your chosen amplicon as a BLAST query to make sure it doesn't bind any other mRNAs. You can, at this step, determine if your probe will bind splice variants and the like. If mRNAs other than your target come up, start over at a different site on the cDNA.

ISH PROTOCOL (DEVELOPED FROM TWO SOURCES[30,31])

1. Cut brain sections fresh or remove them from cryoprotectant. Wash in ice-cold 2× saline-sodium citrate buffer (SSC). If from cryoprotectant, wash three times for 10 min each in 2× SSC or RNase-free PBS (see Chapter 10; RNase-free PBS is only necessary for RNA probes).

   I have used samples from both cryoprotectant and samples from fresh-cut sections. I prefer cryoprotectant, because I can use fewer slices than fresh-cut samples (or else I'm splitting between cryoprotectant and experimental slices and can't save them for later); and I know the tissues are safe in the freezer, while I'm less sure of samples sitting in sucrose solution. Typically, these washes are performed in scintillation vials, with an excess of solution (except prehybridization buffer) to properly wash tissue. The pipetting of solutions using a Pasteur pipette is what really makes this technique take a lot of time. Just make sure you wash in an excess of fluid. Additionally, some researchers use Proteinase K to "free" the mRNA for reaction. Some swear that it works; some swear that it's a waste of time. I avoid proteinases because I've not found them necessary and incubating too long can destroy your tissue!

2. If tissues were fixed with acrolein or glutaraldehyde, incubate tissue for 15 min in 1% sodium borohydride and wash in RNase-free PBS.

   Note that this is only important if you are using HRP to visualize the reaction.

3. Optional: to increase some probe signals, tissue can be incubated in 0.5% Triton X-100 solution at 4°C overnight.
4. Optional: if you are using RNA probes, rinse twice in 0.1 M triethanolamine (TEA; inhibits RNases).
5. Optional: if you are using RNA probes, incubate slices in 0.25% acetic anhydride in 0.1 M TEA for 10 min, followed by three 10 min rinses in 2× SSC.
6. Block in prehybridization buffer for 2 h at 37°C.

7. Add labeled probe to prehybridization buffer (no need to switch in fresh solution) and leave overnight at 37°C.

   Make sure to perform probe titrations as you do with antibodies. You should note that different protocols specify different heats for hybridization, with some as high at 70°C. This may not work well, as high heat can induce artifacts.[32] Determine which heat works well for your experiment. I prefer 37°C, as that is the body temperature of the animal and therefore should work just fine.

8. Rinse sequentially in $2\times$ SSC, $1\times$ SSC, and $0.5\times$ SSC for 10 min each, 37°C.
9. Optional: if using RNA probes, incubate in RNase for 30 min at 37°C.

   This step eliminates any excess probe that did not bind a target, as your RNA probe is, without binding, single stranded and thus a target for RNase. Bound probe-target complexes are double stranded and not targets for RNase.

If using nonradioactive probes, begin staining methods (ABC kit for biotin, anti-DIG antibodies, for example). If using radioactivity, mount on slides, air dry, and expose to X-ray film or emulsion.

ISH CONTROLS[23,33]

Controls for ISH are a little less clear than other techniques. While positive and negative controls are still beneficial, there is a history of using the "sense" strand as a control. The appeal is that it sounds like a control, but it really isn't. The idea is that your sense strand shouldn't bind to anything and is therefore a signal of general, nonspecific binding. What you're really testing there, however, is whether your sense strand is antisense to anything else. As such, the sense control is really meaningless. What should replace it? A "hot/cold" control (named for the slang denoting radioactivity). The idea of a hot/cold control is that there are only so many mRNA strands your probes can bind to; anything extra is nonspecific binding. So you apply an excess of unlabeled ("cold") probe and your typical amount of labeled probe ("hot"). If you see nothing, the cold probe displaced the hot probe at binding sites, and there was nothing left to bind to. If you see "staining," then your probe binds nonspecifically (perhaps specifically as well, but that's cold comfort to you; enjoy the pun).

1. *Positive controls*: as ever, including samples where your target should show up are helpful to determine if there are issues with your fixation or your tissues.
2. *Negative controls*: as ever, including samples where your target mRNA is absent can be helpful to determine specificity of the probe.
3. *Hot/cold probe competition control*: as explained above, tests specificity of the probe.
4. *Multiple probes*: similar to using multiple antibodies, multiple probes against different parts of the mRNA can be compared. If they show similar staining, that suggests that the probes are specific and targeting the same mRNA. Combine with positive and negative controls.
5. *Colocalization with protein*: combining IHC and ISH for the same target can make for a very convincing control. It can be difficult to perform, however, as the heat and chemicals required for hybridization can be incompatible with optimal conditions for IHC.
6. *Northern blots*: another sort of specificity test, though I have no experience with it.
7. *Poly$_{d(T)}$ control*: an offshoot of the normal positive control, poly$_{d(T)}$ binds mRNA tails (all of them, so expect a lot of background staining), and so tests whether you have introduced mRNA degradation to your tissue samples.
8. *Reference gene control*: an offshoot of the normal positive control, this directly tests whether you have introduced mRNA degradation to your tissue samples. This uses a new set of probes but can be very helpful to establish how damaged or undamaged your tissue is.
9. *RNase control*: suggests that your probes are only binding RNA, but this isn't necessarily a great control. RNAs, like proteins, are fixed in three dimensions by your fixative agent (paraformaldehyde, for instance, generates disulfide and other bonds in both proteins and mRNA), and so RNases are a danger to RNA probes but not necessarily to fixed RNA in the tissue.

SOLUTION RECIPES

20× SSC	
NaCl	87.65 g
Citric acid	441 g
DEPC-H$_2$O	To 500 ml
pH to 7.0, autoclave, bring to 500 ml.	

Nuclease-free H$_2$O
See RNA extraction section.

RNase-free PBS
Make 1× or 5× PBS as in the immunohistochemistry section, but add 1 ml DEPC and autoclave.

0.1M TEA	
Triethanolamine (TEA)	1.49 g
DEPC-H$_2$O	To 100 ml
pH to 8.0.	

0.5% Triton X-100	
1 M Tris pH 8.0	5 ml
0.5 M EDTA pH 8.0	5 ml
Triton X-100	250 µl
DEPC-H$_2$O	To 50 ml

Prehybridization buffer	
Formamide	60%
Tris pH 7.4	0.02 M
EDTA	1 mM
Dextran sulfate	10%
Ficoll	0.8%
PVP-40	0.8%
BSA	0.8%
2× SSC	To volume
Dithiothreitol	0.1 M
Herring sperm DNA	1.6 mg/ml

1 ml of prehybridization buffer per vial. Ficoll, PVP, and BSA are together in Denhardt's solution. Tris, EDTA, and dextran can be combined into a TED stock solution.

0.25% acetic anhydride	
Acetic anhydride	10 µl
0.1M TEA	To 4 ml

REFERENCES

[1] Hummon AB, Lim SR, Difilippantonio MJ, Ried T. Isolation and solubilization of proteins after TRIzol extraction of RNA and DNA from patient material following prolonged storage. *Biotechniques* 2007;**42**(4):467−70.

[2] Barker K. *At the bench: a laboratory navigator*. Woodbury, NY, USA: Cold Spring Harbor Laboratory Press; 2004.

[3] Resuehr D, Speiss AN. A real-time polymerase chain reaction-based evaluation of cDNA synthesis priming methods. *Anal Biochem* 2003;**322**(2):287−91.

[4] Ståhlberg A, Håkansson J, Xian X, Semb H, Kubista M. Properties of the reverse transcription reaction in mRNA quantification. *Clin Chem* 2004;**50**(3):509−15.

[5] Brookman-Amissah N, Packer H, Prediger E, Sabel J, editors. *PrimeTime qPCR application guide: experimental overview, protocol, troubleshooting*. 3rd ed. Integrated DNA Technologies; 2012. <http://www.idtdna.com/pages/docs/catalog-product-documentation/primetime-qpcr-application-guide-3rd-ed-.pdf?sfvrsn=11>.

[6] Bustin SA, Benes V, Garson JA, Hellemans J, Huggett J, Kubista M, et al. The MIQE guidelines: minimum information for publication of quantitative real-time PCR experiments. *Clin Chem* 2009;**55**(4):611−22.

[7] Pfaffl MW. A new mathematical model for relative quantification in real-time RT-PCR. *Nucleic Acids* 2001;**29**(9):e45.

[8] Guillemin I, Becker M, Ociepka K, Friauf E, Nothwang HG. A subcellular prefractionation protocol for minute amounts of mammalian cell cultures and tissue. *Proteomics* 2005;**5**(1):35−45.

[9] Schindler J, Jung S, Niedner-Shatteburg G, Friauf E, Nothwang HG. Enrichment of integral membrane proteins from small amounts of brain tissue. *J Neural Transm* 2006;**113**(8):995−1013.

[10] Abcam Protocols Book. Can be ordered or downloaded from the Abcam website; 2012. <http://docs.abcam.com/pdf/misc/abcam-protocols-book-2010.pdf>.

[11] Thermo Scientific. *Western blotting handbook and troubleshooting guide, version 2*. Can be requested from Thermo Fisher website. <http://www.piercenet.com/files/1601959_WB-HB_v3_intl.pdf>.

[12] Gassmann M, Grenacher B, Rohde B, Vogel J. Quantifying Western blots: pitfalls of densitometry. *Electrophoresis* 2009;**30**:1845−55.

[13] Romero-Calvo I, Ocón B, Martínez-Moya P, Suárez MD, Zarzuelo A, Martínez-Augustin O, et al. Reversible Ponceau staining as a loading control alternative to actin in Western blots. *Anal Biochem* 2010;**40**:318−20.

[14] Website of David R. Caprette at Rice University for experimental biosciences and courses. Preparing protein samples for electrophoresis. <http://www.ruf.rice.edu/~bioslabs/studies/sds-page/denature.html>.

[15] Yeung Y-G, Stanley ER. A solution for stripping antibodies from PVDF immunoblots for multiple reprobing. *Anal Biochem* 2009;**389**(1):89−91.

[16] Cell Signaling Technology website. Western blot reprobing protocol. <http://www.cellsignal.com/common/content/content.jsp?id=western-reprobing>.

[17] Thermo Scientific. Tech Tip 24: Optimize antigen and antibody concentrations for Western blots. Available on Thermo website. <http://www.piercenet.com/files/TR0024-Optimize-Ab-dilutions.pdf>.

[18] R & D Systems website. Dot blot protocol. <http://www.rndsystems.com/rnd_page_objectname_wb_dotblot.aspx>.

[19] Abcam website. Dot blot protocol. <http://www.abcam.com/ps/pdf/protocols/Dot%20blot%20 protocol.pdf>.

[20] Cell Signaling Technology website. Western immunoblotting troubleshooting guide. <http://www.cellsignal.com/common/content/content.jsp?id=western-trouble>.

[21] Haan C, Behrmann I. A cost effective non-commercial ECL-solution for Western blot detections yielding strong signals and low background. *J Immunol Methods* 2007;**318**(1−2):11−9.

[22] Thermo Scientific. *Protein interaction technical handbook, version 2.* <http://www.piercenet. com/files/10PIE067_ProteinInteractionHB_1601945.pdf>.

[23] Lal A, Haynes SR, Gorospe M. Clean western blot signals from immunoprecipitated samples. *Mol Cell Probes* 2005;**19**(6):385−8.

[24] Hoffman GE, Le WW, Sita LV. The importance of titrating antibodies for immunocytochemical methods. *Curr Protoc Neurosci* 2008;2.12.1−2.12.26.

[25] LI-COR Biosciences information pack. Good Westerns Gone Bad. 2008. <http://biosupport. licor.com/docs/GoodWesternsGoneBad_11169.pdf>.

[26] Rosebrock JA. *Immunohistochemical double staining: principles and practice.* KPL Website, Tools & Techniques. <http://web.qbi.uq.edu.au/microscopy/cresyl-violet-staining-nissl-staining/>.

[27] Queensland Brain Institute's Advanced Microimaging and Analysis Facility website. Cresyl violet staining. <http://web.qbi.uq.edu.au/microscopy/cresyl-violet-staining-nissl-staining/>.

[28] Wilcox JN. Overview of *in situ* hybridization methodology. Workshop, 2000 meeting of the Histochemical Society. Available at <www.emory.edu/wilcox/>; 2000.

[29] Ishii T, Omura M, Mombaerts P. Protocols for two- and three-color fluorescent RNA *in situ* hybridization of the main and accessory olfactory epithelia in mouse. *J Neurocytol* 2004;**33**:657−69.

[30] Houpt T. wiki.houptlab.org Protocols.

[31] Hoffman G. Personal correspondence.

[32] Moorman AFM, Houweling AC, de Boer PAJ, Christoffels VM. Sensitive nonradioactive detection of mRNA in tissue sections: novel application of the whole-mount *in situ* hybridization protocol. *J Histochem Cytochem* 2001;**49**(1):1−8.

[33] GeneDetect.com website. Laboratory methods section: *in-situ* hybridization using GeneDetect™ oligonucleotide probes. <http://www.genedetect.com/insitu.htm>.

Printed and bound by CPI Group (UK) Ltd, Croydon, CR0 4YY

03/10/2024

01040423-0009